T0275998

CAMBRIDGE LIBRARY COLLECTION

Books of enduring scholarly value

Technology

The focus of this series is engineering, broadly construed. It covers technological innovation from a range of periods and cultures, but centres on the
technological achievements of the industrial era in the West, particularly
in the nineteenth century, as understood by their contemporaries. Infra-
structure is one major focus, covering the building of railways and canals,
bridges and tunnels, land drainage, the laying of submarine cables, and the
construction of docks and lighthouses. Other key topics include developments
in industrial and manufacturing fields such as mining technology, the pro-
duction of iron and steel, the use of steam power, and chemical processes
such as photography and textile dyes.

A Treatise on Mills

From the 1770s onwards, John Banks (1740–1805) lectured on natural
philosophy across the north-west of England. Much of his work aimed
to show engineers, mechanics and artisans how they could benefit from
expanding their theoretical knowledge. First published in 1795, and reissued
here in its 1815 second edition, this work shows how to calculate the power
limits of waterwheels, millstones and other commercially important machines.
In the author's words, a key aim is to avoid wasted effort 'in attempting what
men of science know to be impossible'. Starting with the mechanics of circular
motion, he leads the reader step by step through a series of worked problems,
showing the theory's practical applications. He then moves on to his experiments on the flow of water, and uses his results to better analyse the various
types of waterwheel. Banks' *On the Power of Machines* (1803) is also reissued
in this series.

A Treatise on Mills

In Four Parts

JOHN BANKS

CAMBRIDGE
UNIVERSITY PRESS

CAMBRIDGE
UNIVERSITY PRESS

University Printing House, Cambridge, CB2 8BS, United Kingdom

Cambridge University Press is part of the University of Cambridge.

It furthers the University's mission by disseminating knowledge in the pursuit of
education, learning and research at the highest international levels of excellence.

www.cambridge.org
Information on this title: www.cambridge.org/9781108069830

This edition first published 1815
This digitally printed version 2014

ISBN 978-1-108-06983-0 Paperback

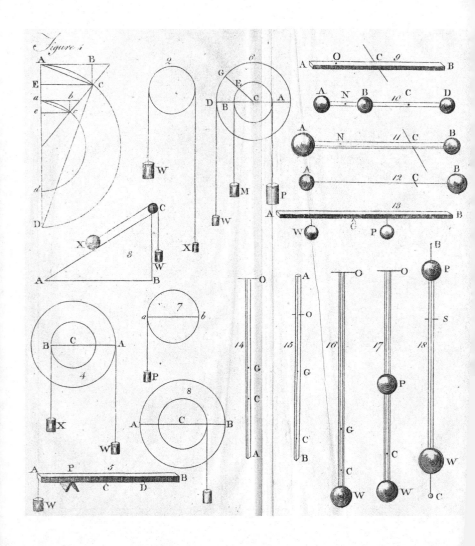

A

Treatise on Mills,

IN FOUR PARTS.

PART FIRST,

ON CIRCULAR MOTION.

PART SECOND,

ON THE MAXIMUM OF MOVING BODIES, MACHINES, ENGINES, &c.

PART THIRD,

ON THE VELOCITY OF EFFLUENT WATER.

PART FOURTH,

EXPERIMENTS ON CIRCULAR MOTION, WATER WHEELS, &c.

BY JOHN BANKS,

LECTURER IN EXPERIMENTAL PHILOSOPHY.

Second Edition.

THEY CONTINUE THIS DAY ACCORDING TO THINE ORDINANCES.
PSALM cxix. 91.

LONDON:

PRINTED FOR LONGMAN, HURST, REES, ORME, AND BROWNE;

AND FOR

W. GRAPEL, LIVERPOOL.

1815.

Printed by G. F. Harris's Widow and Brothers,
Water-street, Liverpool.

Preface.

THE experiments contained in the fourth part of this treatise, and some of these in the third, have been the subject of a public lecture, which I have occasionally delivered, for about twenty years. At the request of many of my hearers, I have made it public; and have prefixed a few problems on circular motion, &c. For in order to know the powers of different wheels, or which is best able to overcome an obstacle, it is necessary to know their central forces.

Where different powers are compared, as in Prob. XII, &c. they are supposed to act upon the wheel, while they descend through the same space. When a wheel or fly, of any given weight or magnitude, has received a given degree of velocity, no more power is required to continue that velocity than what is sufficient to overcome the friction. Yet if it moves twice as fast, it will require four times the former power to continue that motion: hence, one would be apt to infer, that the friction increases with the square of the velocity. In a water-wheel, the same power is not constant; for the same particles act upon the wheel during only a part of a revolution; their places are constantly supplied, or they are succeeded by others which act for their time, &c.: hence, a water-wheel soon acquires an uniform velocity.

In the second part, the third problem has three solutions; the two first on principles the most common, but not perfectly just; but as they are sufficiently near in practice, and much easier than the third, they are continued till the eighth problem, inclusive, which, in fact, is but one problem, for the fourth, &c. are illustrations of the theorems drawn from the third, and ought rather to have been examples. In this, as well as in many of the following problems, I have endeavoured to shew the impossibility of a given power producing more than a given effect, or of raising more than a certain quantity of water, weight, &c. to a given altitude, in a limited time. For, in this case, combinations of machinery afford no assistance, but the contrary. For if one pound in one scale cannot raise one pound in the other, we shall get nothing by suspending a weight from the circumference of a wheel, in order to raise a greater at the circumference of its axle. For if the machine is so constructed that one pound will raise ten pounds, the pound must descend more than ten times as fast as the other rises, and of consequence it must descend more than ten successive times through a given space, to raise ten pounds through the same space.

It is true, that we have in the kingdom many intelligent engineers, and excellent mechanics; and there are others who can execute better than they can design, otherwise there would not have been so much money expended in attempting what men of science know to be impossible.

When a man tells me he can construct a water-wheel in such a manner that, when once put in motion, it shall raise water to keep itself moving; or that he has constructed a pump in such a manner, that one man may do the work of ten, &c. I pay the same attention to him, as if he told me he could create a system of worlds, and command them to move. Or, is he less to be credited who says he can communicate perpetual motion to dead matter? Both are indirectly saying, I can reverse the laws of nature.

Many of the problems in this part are intended not only to prove that a given power can only produce a given effect, but at the same time to demonstrate what the greatest effect is which a given power can produce. For a power is often applied in such a manner as to cause the effect to fall short of the maximum, as demonstrated in the third problem, when the time is given, and in the thirty-fifth, &c. when the space is given, where it appears, that if the power and resistance be equally distant from the centre, in case of a maximum, or when the power produces the greatest possible effect, the former will be to the latter as 1000 to 618. Or if the power be applied ten times as far from the centre as the resistance, when reduced they will be as 1000 to 659, the weight of the machine not considered, for which see Prob. XXXVII, XXXVIII, &c. When the power and resistance are both given, by Prob. XXX, a machine may be constructed, by which the power shall produce the greatest effect in a given time.

In the third part, the experiments on effluent water differ considerably from any theory; and, contrary to expectation, more water is discharged, in proportion.

from small than from large apertures. The quantity
through small holes, I have endeavoured to ascertain,
by different processes, all of which bring out nearly the
same conclusions. And though I have had much practice in making experiments, I have not trusted entirely
to my own observations, but have been assisted in the
whole by one or more gentlemen well acquainted with
the subject. At Coventry, by the Rev. Mr. Banks, of
Monks Kirkby; by Messrs. Baines, Watson, Dicas, &c.
of Liverpool; by Mr. Priestly, and Mr. Peckover, of
Bradford; by my eldest son, and by my wife, who,
though a woman, is perhaps as accurate in making experiments in philosophy, and some branches of chemis-
try, as most of men. Upon the whole, I presume the
rules contained in this part, for finding the quantity of
water discharged in a given time, will be found sufficiently accurate in practice.

However satisfactory mathematical reasoning may
be to some, yet experimental proof is desirable, and, to
many, much more so than the former; and without experiments, we often want *data* to reason from. But if
we have certain principles, the conclusions drawn there-
from will often differ considerably from experiment, or
rather the experiment from the theory. For the theory
supposes the bodies to move in free space, without friction, or any kind of resistance; but as these impediments
cannot be entirely removed, the experiments cannot
perfectly coincide with the theory, though in some cases
they come exceedingly near.

In the fourth part, the experiments on circular
motion, &c. are sufficiently accurate to prove the truth
of the theory, and the utility thereof, when applied to

THE PREFACE.   vii

the construction of machines. I have thought the preceding theory and experiments a necessary introduction to the principal subject, the experiments on wheels, &c.; in which I have endeavoured to investigate the truth, without a view to support any particular system, sentiments, or opinion. And if some of the experiments happen to recommend any construction or application different to the practice of some professional men, it might be well to enquire into the foundation of their practice, whether it is supported by experiments, or whether it rests upon the opinions of their predecessors. To rest satisfied with the opinions of others, however great their reputation, tends much to retard the progress of knowledge; for error is often found in high places.

It will not be expected that I should attempt to instruct the mechanic how to form his cogs, divide the wheels, proportion their diameters, bevels, &c. as numbers of men may be found who can both plan and execute these parts well. The chief end in view is, how to make the most of a given stream; and what the experiments recommend, has long since been put in practice, with acknowledged advantage. If this treatise, in proportion to its sale, proves equally useful with the public lecture, I shall be satisfied.

To conclude: if, through mistake, I have advanced any thing erroneous in theory, or drawn any false conclusions from the experiments, I hope those who discover them will candidly correct them.

JANUARY 29th, 1795.

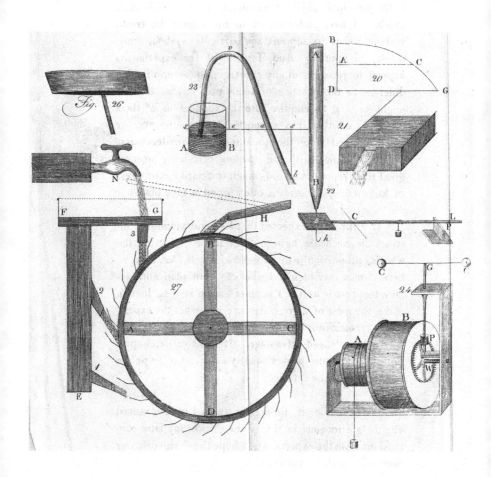

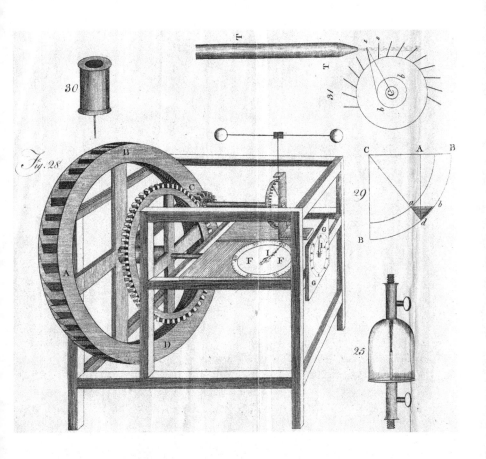

The material originally positioned here is too large for reproduction in this reissue. A PDF can be downloaded from the web address given on page iv of this book, by clicking on 'Resources Available'.

A

Treatise on Mills.

PART I.

OF THE LAWS OF CIRCULAR MOTION, THE RATIOS OF PROJEC-
TILE AND CENTRIFUGAL FORCES, THE PERIODIC TIMES, &c.

———————

FIG. I.

In the larger circle,

D = AD the diameter in feet.

v = AC the projectile force, or orbit velocity in feet.

A = BC = AE the space through which the
body would move towards the centre, while it
i describing AC, or the central force compared
with the projectile.

B

$F = BC$ the centrifugal force compared with gravity, or with the weight of the revolving body, when on the surface of the earth.

$T =$ the periodic time, or time of a revolution, in seconds.

$p =$ the time describing AE or AC.

$s = 16$ feet, the space through which a body falls in 1 second.

$q = 3.1416$, the circumference of a circle, when the diameter is 1.

In the lesser circle,

$d = ad$ the diameter.

$v = ac$ the velocity.

$a = ae$ the central force, &c.

$f = ae = bc$.

$t =$ the time of a revolution.

Every body in motion endeavours to move in a straight line; the force which causes it to leave that line is called the *centripetal*, and the resistance which it affords the *centrifugal* force.

1. If a body at A, moving towards B, is drawn to C, BC represents the centrifugal *force*, and AE the centripetal *force*, which are equal; and as AE is to AC, so is the centrifugal to the projectile force, or circular velocity.

2. From the property of the circle,

As AD : AC :: AC : AE $=\frac{\overline{AC^2}}{AD}=$ A.

Or, As D : v :: v $:\frac{v^2}{D}=$ A.

3. And in equable motion it will be,

As the time of a revolution is to the circumference of the orbit, so is any other time to the space passed over in that time; or so is 1 second to the velocity per second.

Viz. As T : Dq :: p $:\frac{Dqp}{T}=$ v.

And by substituting $\frac{Dqp}{T}$ for its equal AC or v

we have, As D $:\frac{Dqp}{T}::\frac{Dqp}{T}:\frac{Dq^2p^2}{T^2}=$ AE $=$ A.

Or, As D : v :: v $:\frac{v^2}{D}=$ AE $=$ A.

4. Also, As $1^{2''}$: s :: p^2 : sp^2, the space through which a body near the surface of the earth would fall in the time p.

5. Then, to compare the centrifugal force with the gravity, or weight of the revolving body, which put $=1$.

It will be, As sp^2: AE :: I $:\frac{AE}{sp^2}=$ F.

Or, As $sp^2:\frac{Dq^2p^2}{T^2}::$ I $:\frac{Dq^2}{sT^2}=$ F.

Or, As $sp^2:\frac{v^{2'}}{D}::$ I $:\frac{v^2}{sDp^2}=$ F.

When the central force is equal to the weight of the body, $\dfrac{Dq^2}{ST^2} = 1$, and $sp^2 = 16$, if $p = 1$ second; also $A = s$.

From these proportions we obtain the following Theorems:

Theo. 1. $A = \dfrac{v^2}{D} = \dfrac{vq}{T} = \dfrac{Dq^2}{T^2} = SF.$

Theo. 2. $v = \sqrt{AD} = \dfrac{Dq}{T} = \sqrt{FDS}.$

Theo. 3. $T = \dfrac{vq}{A} = \dfrac{Dq}{v} = q\sqrt{\dfrac{D}{A}} = q\sqrt{\dfrac{D}{SF}} = \dfrac{vq}{SF}.$

Theo. 4. $D = \dfrac{v^2}{A} = \dfrac{v^2}{SF} = \dfrac{T^2A}{q^2} = \dfrac{SFT^2}{q^2}.$

Theo. 5. $F = \dfrac{Dq^2}{ST^2} = \dfrac{A^2D}{Sv^2} = \dfrac{v^2}{SD} = \dfrac{A}{S}.$

6. If two bodies, in different circles, revolve in the same time, the velocities will be directly as the diameters of the circles. $(T = t.)$

For, As $Dq : v \ (AC) :: dq : v \ (ac) = \dfrac{vd}{D}$;

and, As $D : d :: v : v = \dfrac{vd}{D.}$ Q. E. D.

And, As $AC : AE :: ac : ae.$

Or, As $v : A :: v : a = \dfrac{Av}{v.}$

Or, As $v : F :: v : \dfrac{vF}{} = f.$

Hence, As $v : v :: F : f :: D : d.$

From which it appears that, if the times are equal, the central force is directly as the diameter.

7. If $v = v$, then $Dq : dq :: T : \frac{Td}{D} = t.$

Therefore, As $D : d :: T : t.$

And, As $\frac{D}{T} : \frac{d}{t} : v : v = \frac{vdT}{Dt.}$

$\frac{D}{T} : \frac{d}{t} :: F : f = \frac{FdT}{Dt.}$ and because F is as $\frac{v^2}{D}$ we

have, As $F : f :: \frac{v^2}{D} : \frac{v^2}{d}.$ Or,

As $DF : df :: v^2 : v^2,$ and $v : v :: \sqrt{DF} : \sqrt{df}.$

And $\frac{D}{T}$ being as v, we have $\frac{D}{T}\sqrt{df} = \frac{d}{t}\sqrt{DF}.$

Or, from the fifth Theorem, if $T = t.$

As $\frac{Dq^2}{ST^2} : \frac{dq^2}{ST^2} :: F : f = \frac{Fd}{D.}$

Theo. 6. $v = \frac{vd}{D.}$

Theo. 7. $d = \frac{vD}{V.}$

Theo. 8. $a = \frac{v^2d}{D.}$

Theo. 9. $f = \frac{v^2d}{SD.}$

8. If the diameter remains the same; or if $D = d,$

Then, As $\frac{Dq^2}{sT^2}:\frac{Dq^2}{sT^2}::F:\frac{FT^2}{t^2}=f$, and as $t^2:$
$T^2::F:f::v^2:v^2$, or the central forces, in the same circle, are reciprocally as the squares of the times; or directly as the squares of the velocities.

Theo. 10. $f=\frac{FT^2}{t^2}$.

Theo. 11. $t=\sqrt{\frac{FT^2}{f}}$.

9. When $F=f$; $\frac{Dq^2}{sT^2}=\frac{dq^2}{st^2}$, or $Dt^2=dT^2$.

And $\frac{D}{T^2}=\frac{d}{t^2}$ (see Theo. 5.) $\frac{Dt^2}{T^2}=d$.

And, As $T^2:t^2::D:d$, or $T:t::\sqrt{D}:\sqrt{d}$.

Theo. 12. $d=\frac{Dt^2}{T^2}$ when the times are given.

Theo. 13. $t=T\sqrt{\frac{d}{D}}$ when the distances are given.

The foregoing Theory exemplified, in the solution of various Problems in circular motion.

PROBLEM I.

Given the diameter of the orbit, 10 feet, and the centrifugal force equal to the weight of the revolving body; required the time of a revolution, and velocity per second?

In this case, $A = s$; and $F = i$ (see **Art. 5.**)

In Theo. 3. we have $T = q\sqrt{\frac{D}{SF}} = 3.1416\sqrt{\frac{10}{16}}$
$= 2.4818''$; and per Theo. 2. $v = \frac{Dq}{T} = \frac{31.416}{2.4818}$
$= \sqrt{FDS} = \sqrt{160} = 12.6491$.

PROBLEM II.

Given the time of a revolution, 3 seconds, the central force, 1; required the distance, and velocity?

$\left. \begin{array}{l} T = 3. \\ F = 1. \end{array} \right\}$ Theo. 4. $D = \frac{T^2 s}{q^2} = \frac{144}{9.8696} = 14.59$

feet; half of which is the distance, $= 7.295$ feet.

And $v = \sqrt{FDS} = 15 \cdot 279$ feet per second.

PROBLEM III.

Given the diameter and periodic time, to find the central force and velocity?

$\left. \begin{array}{l} D = 14.59. \\ T = 3''. \end{array} \right\}$ (Theo. 5.) $F = \frac{Dq^2}{ST^2} = \frac{144}{144} = 1.$

$v = \frac{Dq}{T} = \frac{45.837}{3} = 15.279$ feet.

8 PART FIRST.

PROBLEM IV.

Given the diameter, 14.59 feet, and the velocity, 15.279, to find the periodic time and central force?

(Per Theo. 3.) $T = \dfrac{Dq}{v} = \dfrac{45.837}{15.279} = 3''$.

(Per Theo. 5.) $F = \dfrac{v^2}{sD} = \dfrac{233.44}{16 \times 14.59} = 1$.

PROBLEM V.

Given the diameter, 14.59 feet, the central force equal to twice the weight of the body; what is the velocity, and time of a revolution?

$T = q\sqrt{\dfrac{D}{2Fs}} = \sqrt{\dfrac{14.59}{32}} \times q = .6753 \times 9 = 2\cdot1214''$.

$v = \sqrt{2FDs} = 21.6074$ feet per second.

Or if we compare this problem with the last, we have (8)

As $f : F :: T^2 : t^2$, or $t = T\sqrt{\dfrac{F}{f}} = 3\sqrt{\dfrac{1}{2}} = 2.12$, &c.

And, As $F : f :. v^2 : v^2 = \dfrac{v^2 f}{F}$, or $v = v\sqrt{\dfrac{f}{F}} = 15.279 \times \sqrt{\dfrac{2}{1}} = 15.279 \times 1.414$, &c. $= 21.607$.

PROBLEM VI.

Let the diameter be 29.18 feet, (twice as much as in Prob. III.) the time of a revolution 3 seconds; required the velocity, and central force?

$$v = \frac{Dq}{T} = \frac{91.671}{3} = 30.5572.$$

$$F = \frac{Dq^2}{ST^2} = \frac{288}{144} = 2.$$

(Art. 6.) As $D : d :: V : v = \frac{vd}{D} = 30.5572.$

Viz. As $14.59 : 29.18 :: 15.279 : 30.5572.$

Again, As $D : d :: F : f = \frac{dF}{D}.$

As $14.59 : 29.18 :: 1 : 2$, the centrifugal force, the same as above.

PROBLEM VII.

The stones on which they grind table knives at Sheffield, are about 44 inches diameter, and weigh about half a ton; the velocity of the surface is at the rate of 1250 yards in a minute, equal to 326 revolutions; required the centrifugal force, or the tendency which the stones have to burst?

$D = 2.59$ feet, the diameter of the circle of gyration.

c

$\text{T} = .184$ seconds, the time of one revolution.

$\text{T}^2 = .033856.$

$$\text{F} = \frac{\text{D}q^2}{\text{S}\text{T}^2} = \frac{2.59 \times 9.8696}{16 \times .0338,\&c.} = \frac{25.5622}{.54169} = 47.18$$

times the weight of the stone, or $23\frac{1}{2}$ tons.

PROBLEM VIII.

If a fly, 12 feet diameter, and 3 tons weight, revolves in 8 seconds, and another of the same weight revolves in 3 seconds; what must be the diameter of the last, when they have the same centrifugal force?

$\text{D} = 12.$

$\text{T} = 8.$

$t = 3.$

$\text{F} = f.$

And per Theo. 12. $d = \frac{\text{D}t^2}{\text{T}^2} = \frac{108}{64} = 1.6875$

feet, the diameter of the circle of percussion.

N. B. As the weight of each fly is the same, it does not enter into the solution; but if the diameters should be the same, as in the next Problem, the weight must be considered.

PROBLEM IX.

If a fly, 12 feet diameter, revolves in 8 seconds, and another of the same diameter in 3 seconds; what are the ratios of their weights, when the central forces are equal?

Here, $F = f$, and $D = d$, which put $= 1$.

And (Art. 8.) when $D = d$, we have, As t^2: $T^2 :: F : f$; As $9 : 64 :: 1 : \dfrac{64}{9} = 7\frac{1}{9}$, the centrifugal force of the second, when they are of equal weight.

And when the distance and time are the same, the force is directly as the weight; the weight of the second will therefore be obtained by dividing the weight of the first by $7\frac{1}{9}$.

PROBLEM X.

If a fly, 2 tons weight, and 16 feet diameter, is sufficient to regulate an engine, when it revolves in 4 seconds; what must be the weight of one 12 feet diameter, when it revolves in 2 seconds, so that it may have the same power upon the engine?

Here we have $D = 16$; $d = 12$; $T = 4$; $t = 2$; $F = f$; suppose d unknown, then, by Theo. 12, $d = \dfrac{Dt^2}{T^2} = 4$, when the weights are equal, viz. if the first is 16 feet diameter, and the second 4 feet, their centrifugal forces would be equal; but, by the question, the second is to be 12 feet diameter, therefore, (whether it revolves in a circle of 4 feet or 12 feet, the time is 2 seconds) As $D : d :: F : f$, (Art. 6.) viz. As $4 : 12 :: 1 : 3 = f$, but $f = F$ by the question, hence the central force, and. of consequence, the weight is three times too great, and the second fly ought to be one-third of the weight of the first. $= \dfrac{40}{3} = 13\frac{1}{3}$ cwt.

Or, the two last Problems may be solved otherwise.

In Prob. IX. Let $w =$ the weight of the first fly $= 3$ tons. $x =$ the weight of the second.

And (Art. 8.) we have $t^2 w = T^2 x$, and $x = \dfrac{t^2 w}{T^2}$ $= \dfrac{27}{64} = .4218 = 8.43$ cwt. $= \dfrac{60}{7\frac{1}{9}}$, as before.

In Prob. X. we have $\dfrac{D t^2 w}{d T^2} = x = \dfrac{128}{192} = .6666$ $= 13\frac{1}{3}$ cwt.

PROBLEM XI.

If a cast-iron fly, 12 feet diameter, and 2 tons weight, has a sufficient centrifugal force to regulate the pin shaft of a forge-hammer, when it revolves in 2 seconds; required the weight of a fly, of the same diameter, that shall have the same force, when it revolves in 4 seconds?

$T = 2$; $t = 4$; $D = d$; $F = f$; and $w = 2$; hence, $\dfrac{t^2 w}{T^2} = x =$ the weight sought $= \dfrac{16 \times 2}{4} =$ 8 tons, or four times the weight of the first.

N. B. When the weight of the revolving body and time of a revolution remain the same, the central force is directly as the distance from the centre, or as the diameter of the wheel; (see Theo. 5.) but if the weight increases with the distance, or diameter, (and in water-wheels it often increases in a greater ratio) then the centrifugal force will be as the square of the diameter, viz. F is as $D^2 \dfrac{D^2}{T^2}$, or as $\dfrac{D^2 q^2}{s T^2}$.

PROBLEM XII.

Let the centrifugal force of a wheel 6 feet diameter, and that of another of 12 feet diameter, be equal; or, in other words, let the impelling power be the same, and the first revolve in 5 seconds; required the time in which the second revolves?

Here $F = f$; $D = 6$; $d = 12$; and $T = 5$.

By comparing Theo. 13. with the note to
Prob. XI. we get $t = \dfrac{\text{T}d}{\text{D}} = 10$ seconds.

PROBLEM XIII.

Let the two wheels revolve in the same time;
required the ratio of the impelling powers, or
the central forces?

(See Art. 6.) As $\text{D}^2 : d^2 :: \text{F} : f = \dfrac{\text{F}d^2}{\text{D}^2}$.
Let $\text{F} = 1$ then $\dfrac{\text{F}d^2}{\text{D}^2} = \dfrac{144}{36} = 4$.

Viz. the 12 feet wheel requires four times as
much power as the 6 feet wheel, to turn in the
same time.

Otherwise, $\text{F} = \dfrac{\text{D}^2}{\text{T}^2}$; $f = \dfrac{d^2}{t^2}$; but, As $\text{T} = t$.
$\text{F} = \text{D}^2$, and $f = d^2$; Or,
As $\text{F} : f :: 36 : 144 :: 1 : 4$ as before.

PROBLEM XIV.

Given the diameter of a water-wheel, 12 feet,
its weight 2 tons, the time of a revolution 9
seconds; the diameter of the cog-wheel 10
feet, weight 12 cwt.; diameter of mill-stone 6
feet, weight 1 ton, time of a revolution six-
tenths of a second; required the centrifugal

force of the water wheel, cog-wheel, and mill-stone?

In the water-wheel, $D = 12$; $T = 9$; $w = 40$ cwt. And $\dfrac{Dq^2}{ST^2} \times w = F = \dfrac{4733}{1296} = 3.65$ cwt.

In the cog-wheel, $D = 10$; $T = 9$; $w = 12$. And $\dfrac{Dq^2w}{ST^2} = F = \dfrac{1183}{1296} = .91$.

In the mill-stone, $D = 4$ (the diameter of the circle of percussion;) $T = .6$; $w = 20$.   And $\dfrac{Dq^2w}{ST^2} = F = \dfrac{789}{5.76} = 137$ cwt.

Water-wheel - - - - - - - - -   3.65
Cog-wheel - - - - - - - - - - -   .91
Mill-stone - - - - - - - - - - - 137.00

Centrifugal force of the whole - 141.56 $=$ 7 tons 1.56 cwt.

In the above computation, it is supposed that the trundle, which turns the stone, is itself turned by the cog-wheel; its diameter therefore could not be more than 9 inches, for which reason it is not taken into the computation.

PROBLEM XV.

Let the water-wheel and cog-wheel be the same as in the last Problem, the diameter of the trundle 2.5 feet, its weight 2 cwt, and on the same axis a cog or spur-wheel 10 feet diameter, and 12 cwt., which turns two trundles each 2.5 feet diameter, and weight together 4 cwt., the stones turned by them each 5 feet diameter, and weight together 36 cwt.; required the centrifugal force of the whole?

The central force of the water and cog-wheel, as in the last Problem, is $= 4.56$ cwt.

The diameter of the cog-wheel, being divided by that of the trundle, quotes 4; the number of revolutions that the trundle makes for one of the water-wheel, $=$ one turn in $2\frac{1}{4}$ seconds.

And we may take $\text{D} = \frac{2}{3}$ of $2.5 = 1.66$; $\text{T} = 2.25$; $w = 2$. And $\frac{\text{D}q^2w}{\text{S}\text{T}^2} = \text{F} = \frac{32.774}{81} = .404$ cwt. the force of the trundle.

In the spur-wheel, $\text{D} = 10$; $\text{T} = 2.25$; $w = 12$. And $\frac{\text{D}q^2}{\text{S}\text{T}^2} \times w = \frac{1183.2}{81} = 14.6$ cwt. the force of the spur-wheel.

The second trundle, or nut, will revolve four times for the spur-wheel once; therefore, if we divide 2.25, the time in which the latter

revolves, by 4, we shall have the time in which
the trundle and stone revolve, $=.5625$ second.

$D = 1.66$; $T = .5625$; $w = 2$. And $\frac{Dq^2w}{4sT^2} =$
$F = 3.277$, which doubled, gives 6.554, the
force of the two trundles.

Lastly, for the force of the stones, we have
$D = 3.33$; $T = .5625$; $w = 18$. And $\frac{Dq^2}{sT^2} =$
$F = 117$, which, multiplied by 2, gives the centrifugal force of both mill-stones, $= 234$ cwt.

Force of the water & cog-wheel   4.56
First trundle - - - - - - - - -   .404
Spur-wheel - - - - - - - - - - 14.6
Second trundles - - - - - - - - 6.554
Mill-stones - - - - - - - - - -234.
                                 ————
Central force of the whole - -250.118 cwt.

It may be observed, that the power which
turns the wheel, or wheels, when applied in
the same manner, has always the same ratio to
the computed central force; and if a given
power produces the forces computed in this Problem, the ratio of the power to turn it round,
in any other time, is easily obtained, as follows:

PROBLEM XVI.

What power, compared with the last, would
turn the wheel in 12 seconds?

Per Theo. 10. $t^2 : \text{T}^2 :: \text{F} : f = \frac{\text{FT}^2}{t^2}$.

Viz. As $12^2 : 9^2 :: 250 : 140$ nearly, or for every 250lb. in the first case. 140lb. will be sufficient in the second.

Observations on the quantity of water required to produce different velocities in the same wheel, or train of wheels.

It has been demonstrated, (see Theo. 5 and 9, and Art. 8.) that in the same wheel, or wheels, the force is directly as the square of the velocity: viz. if a weight, suspended by a cord coiled round the axis of a wheel, can turn that wheel 10 times in a minute, it will require four times as much weight to turn it 20 times in one minute, or to communicate a double velocity. But if the power is a given stream of water, the case will be different.

Suppose a wheel is turned ten times in a minute, and that every bucket receives a gallon of water, (or any other quantity) if it is turned 20 times in a minute, a bucket can receive but half a gallon. Hence the stream must be doubled to supply every bucket with a gallon. But they ought to receive four gallons each. Therefore the stream, when doubled, must be multiplied by 4, or made eight times as large as at first, to supply the buckets with four times the quantity; consequently the velocity of the wheel will be as the cube root of the quantity of water, when it acts by gravity.

PROBLEM XVII.

If a given quantity of water turns the drum, &c. of a cotton-mill 50 times in a minute ; what quantity will be required to turn it 40 times in a minute?

Here the velocities are as the number of turns made in a given time, viz. as 50 to 40, or as 5 to 4 ; the cubes of which are as the quantities of water which produce them, viz. $5 \times 5 \times 5 = 125$, and $4 \times 4 \times 4 = 64$. Hence, for every 125 hogsheads in the first case, 64 will be sufficient in the second.

PROBLEM XVIII.

If a given quantity of water turns a wheel six times in a minute; how often will it be turned by double the quantity?

The velocities being as the cube root of the quantities, or as $\sqrt[3]{1} : \sqrt[3]{2}$, viz. as $1 : 1.2599$: and as one degree of velocity is to six turns, so is 1.2599 to 7.5594 per minute, the number of turns sought.

A

Treatise on Mills.

PART II.

TO INVESTIGATE THE VELOCITY OF A MACHINE; THE RATIO OF
THE POWER AND RESISTANCE BEING GIVEN.

PROBLEM I.

FIG. II.

LET a weight a be suspended at one end of a cord passing over a wheel, and a weight, or resistance, x, at the other, to determine the space passed over in a given time.

It is evident, that $a + x$ is the quantity of matter to be moved, and that $a - x$ is the moving power.

EXAMPLE.

Let $a = 10$; $x = 5$; $s = 16$ feet, the space fallen through by a heavy body in one second; $v =$ the space sought.

Not taking the weight of the wheel into the expression, we have,

As $a + x : a - x :: s : \frac{sa - sx}{a + x} = v =$ the space passed over in the first second of time by the moving bodies, $= \frac{16 \times 10 - 16 \times 5}{10 + 5} = 5\frac{1}{3}$ feet.

PROBLEM II.

FIG. III.

Let w and x be two bodies connected by a cord ; w hanging perpendicular, x supported by the inclined plane ; required the accelerating force, and the ratio of x to w, when the effect is the greatest?

Put $1 =$ radius.
$s =$ sine of the angle of elevation.

Then as the radius (1) is to (s) the sine of the angle of elevation, so is x to sx, the force with which the weight x endeavours to descend down the plane.

And as the sum of their forces is to the difference, so is the force of gravity to the accelerating force.

Viz. As $w + sx : w - sx :: 1$ (the force of gravity) : $\frac{w - sx}{w + sx}$: the space passed over in a given time, compared with gravity, or with the space passed over by a falling body.

The greatest effect of w will be when the momentum of x is a maximum, which will be had by multiplying the velocity by the quantity.

Viz. $\frac{w - sx}{w + sx} \times x = $ maximum.

Or, $\frac{wx - sx^2}{w + sx} = $ the momentum, and must be a maximum by the question; in fluxions

$$\overline{w\dot{x} - 2sx\dot{x}} \times \overline{w + sx} - sx \times \overline{wx - sx^2} = 0,$$

from which we get $s^2 x^2 + 2swx = w^2$; by completing the square, and extracting the root, we have $x = \frac{w}{s}\sqrt{2} - \frac{w}{s}$.

EXAMPLE.

Let $w = 10$; $s = .5$ the natural sine of 30°, to find x?

$\frac{w}{s}\sqrt{2} = 28.28$, from which substract $\frac{w}{s}$ (20)
and we have $8.28 = x$. In case of a pully,
$s = 1$, and $x = w\sqrt{2} - w$.

PROBLEM III.

FIG. IV.

Given a weight w, suspended from a wheel
A; required the weight x, suspended from the
axis B, when the effect is a maximum?

Let BC $= a = 1$; AC $= b$; $s = 16$ feet;
the velocity of $x = 1$; and as $a : b :: 1 : \frac{b}{a}$, and
when $a = 1$, the velocity of $w = b$, and (per
mechanics) the momentum of w will be $= wb$,
and that of $x = a$, or x, when $a = 1$; therefore
$wb + x =$ the sum of the momentums, and as
$wb + x : wb - x :: s : \frac{swb - sx}{wb + x} = v$, the velocity of w; and as AC : BC :: $b : a ::$ the velocity
of w: the velocity of $x = \frac{swb - sx}{wb^2 + bx}$, which multiplied by x gives its momentum, $\frac{swbx - sx^2}{wb^2 + bx}$,
which must be a maximum.

In fluxions, $\overline{bwx - 2xx} \times \overline{wb^2 + bx} - b\dot{x} \times$
$\overline{wbx - x^2} = 0$, and $b^2w^2 - bwx - 2x^2 = bwx$
$- x^2$; from which we find $x = bw \times \sqrt{2} -$

bw: viz. if *bw* is multiplied by the square root of 2 (1.414214) and *bw* taken from the product, the remainder is *x*; or if 1 is taken from the square root of 2, and the remainder (.414214) is multiplied by *bw*, the product is *x*.

NOTE.—Whether we multiply the velocity of *w*, or the velocity of *x*, by *x*, to obtain the momentum, the value of *x* will be the same.

Otherwise, Let $v =$ the velocity of *x*, then As $a : v :: b : \dfrac{bv}{a} =$ the velocity of $w = bv$ when $a = 1$, and $\dfrac{x}{b} =$ the power which applied at B will balance *x*, hence $\dfrac{bw-x}{b}$ is the accelerating force $=f$; and $vx + \dfrac{x}{b} + f \times bv =$ the sum of the momentums $= f = \dfrac{bw-x}{b}$, viz. $2xv + bfv = \dfrac{bw-x}{b}$; and by substituting $\dfrac{bw-x}{b}$ for *f*, we have $vx + bwv = \dfrac{bw-x}{b}$, and $v = \dfrac{bw-x}{b^2w+bx}$; the same as before.

Again, the same notation remaining, $\dfrac{ax}{b} =$ the power which balances *x*, and $\dfrac{bw-x}{b} =$ the accelerating force at *x*, and $\dfrac{a^2x}{b^2} + w =$ the whole resistance reduced to B, by which divide $\dfrac{bw-x}{b}$

and we have $\dfrac{b^2w - bax}{b^2w + a^2x}$ for the force at B, which multiplied by x, gives $\dfrac{b^2wx - bax^2}{b^2w + a^2x}$ for the momentum.

The fluxion of which is $\overline{b^2w\dot{x} - 2bax\dot{x}} \times$ $\overline{b^2w + a^2x} - a^2\dot{x} \times \overline{b^2wx - bax^2} = 0$ and $x = \dfrac{bw}{a}\sqrt{\dfrac{b^2 + ba}{a^3}} - \dfrac{b^2w}{a^2}$; or if $a = 1$, $x = bw\sqrt{b^2 + b} - b^2w$.

Mathematicians in general have estimated the forces of moving bodies by multiplying the quantity into the velocity, as in the first and second solutions of this Problem. The last solution is on different principles, except in finding the momentum, where the accelerating force is multiplied by the quantity; and though it gives the value of x exactly the same when a and b are equal, yet it varies a little when they are unequal. And the effect of w, although a maximum, will vary with the ratio of $a : b$: for instance, if $w = 10$, $a = 1$, $b = 2$, the momentum of x will be 2.0204; but if $b = 10$, the effect will be 2.382, and if $b = 100$, x will be found 498, and its momentum 2.487.

If any three of the four terms a, b, w, x, be given, the fourth may be found, when the effect is the greatest, by the following Theorems:

E

Let $n = \sqrt{2} - 1 = .414214$.

Then. Theo. 1. $x = \dfrac{bwn}{a}$.

Theo. 2. $w = \dfrac{ax}{bn}$.

Theo. 3. $a = \dfrac{bwn}{x}$.

Theo. 4. $b = \dfrac{ax}{wn}$.

The two following Theorems give the velocities, whether the effect be a maximum or otherwise.

Theo. 5. $v = \dfrac{wb - xa}{wb + xa} = $ velocity of w.

Theo. 6. $v = \dfrac{wb - x}{wb + x} \times \dfrac{a}{b} = $ velocity of x.

If the velocities thus found be multiplied by 16, the product is the number of feet passed over in the first second.

PROBLEM IV.

Given a, b, and w, to find x?

Let $a = 2$; $b = 3$; $w = 10$. Then,

(Theo. 1.) $x = \dfrac{bwn}{a} = \dfrac{3 \times 10 \times .4142}{2} = 6.21321$.

PROBLEM V.

Given a, b, and x to find w?

Let $a = 10'$; $b = 12$; and $x = 20$.|

(Theo. 2.) $w = \dfrac{ax}{bn} = \dfrac{200}{4.9705} = 40.23$.

PROBLEM VI.

Given $a = 1$, $x = 100$, $w = 20$; required b, the distance of w?

(Theo. 4.) $b = \dfrac{ax}{wn} = \dfrac{100}{8.2842} = 12.071$.

PROBLEM VII.

Given $w = 100$, $b = 10$, $x = 500$; required a, the distance of x?

(Theo. 3.) $a = \dfrac{bwn}{x} = .82842$.

PROBLEM VIII.

The same terms given, to find the velocity of w?

(Theo. 5.) $v = \dfrac{wb - ax}{wb + ax} = \dfrac{1000 - 414.21}{1000 + 414.21} =$
$\dfrac{585.89}{1414.21} = .414 \times 16 = 6.6274$ feet the first second.

In the following Table, the power w remains the same; the weight x varies with the distance of w, and is found by the foregoing Theorems; and the same power in every example produces the same momentum, viz. the weight x, multiplied by its velocity, gives 27.4516, which it can never exceed.

TABLE I.

No.	a	x	b	w	Velocity of x	Momentum of x.
1	1	4.142	1	10	6.6274)	27.4516+
2	1	8.284	2	10	3.3137	27.4516
3	1	12.426	3	10	2.2091	27.4516
4	1	16.568	4	10	1.6568	27.4516
5	1	41.42	10	10	.6627	27.4516
6	1	82.84	20	10	.3313	27.4516

For if the distance of w is increased, the momentum of x will be diminished.

For example, let a, x, and w be the same as in No. 1 of the Table, and let $b = 2$; then $v = \frac{wb - x}{wb + x} = 10.5088$, which divided by b, gives 5.2544, the velocity of x, which multiplied by x, gives the momentum, $= 21.7637$, but in the Table it is 27.45.

Let a, b, and x be varied as in Table II. while the power w remains the same; the true momentum, computed from the given term, is always less than the maximum, except No. 2.

TABLE II.

No.	a	x	b	w	Velocity of x.	Momen-tum of x.	Velocity of w.
1	1	10	2	10	2.666	26.66	5.333
2	1	10	2.4142	10	2.74516	27.4516	6.627
3	1	10	4	10	2.4	24	9.6
4	1	10	50	10	.30744	3.0744	15.372
5	1	10	1000	10	.01596	.1596	15.96
6	0	0	10	10	.0000	.000	16

The greatest space that w can possibly pass over in the first second is 16 feet; but if it has any resistance to overcome, it must fall short of that space, and that more or less as the circumstances may be: and if the resistance of the machine is not taken into consideration, the effect is a maximum, or the greatest, *when the momentum of the power is to the momentum of the weight as* 10 *to* 4.14214; *or as* 70 *to* 29, *nearly.*

In the above Problems, the accelerating force compared with gravity, always is as the difference of the weights to their sum, or as the difference divided by the sum.

But if the moving power, or force, acted according to any different law, the ratio of the power to the weight, in case of a maximum, would vary also. As for instance, where a wheel is impelled by the force of a given stream, and which force varies with the velocity of the wheel, viz. it is always as the square of the velocity with which it strikes the wheel, as demonstrated in the next Problem.

PROBLEM IX.

Suppose the power to be the impulse of a stream of water upon the floats of a wheel.

Let the quantity of water be represented by 1, and the velocity by 1, these multiplied together give the momentum; let the velocity be doubled, and it follows that the quantity will be doubled, for while the aperture remains the same, a double velocity must discharge a double quantity: hence, the momentum will always be as the square of the velocity.

Let v = velocity of the stream.

x = velocity of the wheel.

$v - x$ = velocity with which the stream stikes the wheel.

And $\overline{v-x}|^2$ will be the force with which the stream impels the wheel; this multiplied by the velocity of the wheel, gives the momentum thereof; and when the effect is greatest it must be a maximum; hence, $\overline{v - x}|^2 \times x = $ M, a maximum.

And $v^2\dot{x} - 4v x \dot{x} + 3x^2\dot{x} = 0$.

$x^2 - \dfrac{4vx}{3} = -\dfrac{v^2}{3}$; from which we get $x = \dfrac{v}{3}$.

And as $v^2 : \overline{v-x}|^2 :: 1 : \dfrac{\overline{v-x}|^2}{v^2} :: 9 : 4$.

EXAMPLE.

Let $v = 6$; then $x = 2$; and $v - x = 4$.

And as $v^2\ (36) :: \overline{v-x}|^2\ (16) :: 1 : \dfrac{4}{9}$.

Hence it appears, that when the velocity of a wheel, which is driven by impulse only, is one-third the velocity of the stream, the effect is the greatest, and the power is to the resistance as 9 to 4.

In the above process the weight of the wheel is not considered; hence, the force acting upon it may be considered as the weight, which multiplied by the velocity, gives the relative effect, or momentum; and the conclusion is the same as if w had been taken to represent a weight, which, suspended from the wheel, would balance the impulse of the water.

In order to estimate the force of a stream upon an undershot wheel, it has been common to take the area of the section, and depth; but this method is exceedingly erroneous, (as will

appear in the third part of this work) and perhaps we cannot, from any known principles,
independent of experiments, compute the true
force of a striking fluid; we must therefore
proceed with the relative force, which is as
the square of the *velocity*, while the aperture
remains the same; but if the depth remains
the same, and the aperture varies, the force is
directly as the area of the aperture.

In this Problem the effect produced is as
$\overline{v-x}|^2 x$; let $v = 6$, $x = 1$.

Then $\overline{v-x}|^2 x$　　$=$　　$25.$ ⎫ The rela-
$x = 2.$ Then $\overline{v-x}|^2 x = 32.$ ⎪ tive effects of
$x = 3.$ Then $\overline{v-x}|^2 x = 27.$ ⎬ the different
$x = 4.$ Then $\overline{v-x}|^2 x = 16.$ ⎪ velocities of
$x = 5.$ Then $\overline{v-x}|^2 x = 5.$ ⎭ the wheel.

PROBLEM X.

FIG. V.

Given a lever equally thick, to find the
length, PB, so that it may, by its own weight,
just balance the end AP, and weight w?

Let $z =$ AP.
$x =$ PB, the length sought.
$m =$ the weight of one foot in length of the
lever.

Then, $zm =$ the weight of AP ; and $xm =$ the weight of PB : but by the question, the momentum of PB must be equal to the momentum of AP $+$ w.

The momentum of PB is mxx, that is, the weight multiplied by the fluxion of the length; in like manner, the momentum of AP is mzz, and the momentum of w is wz : hence, $mxx = mzz + wz$.

The fluent of which is $\frac{mx^2}{2} = \frac{mz^2}{2} + wz$; and $x^2 = z^2 + \frac{2wz}{m}$; and $x = \sqrt{z^2 + \frac{2wz}{m}}$.

Otherwise, without fluxions, let $x =$ PB, the length sought, as before, and make PC $=$ AP $= z$; and PD $= (\frac{CB}{2} + PC)\frac{x+z}{2} =$ the distance of the centre of gravity of CB from the prop P, which multiplied by its weight, $mx - mz$, gives $\frac{mx^2 - mz^2}{2}$ for the momentum, which must be equal to zw ; the momentum of AP is equal to that of PC, and therefore not considered ; hence we have, $mx^2 - mz^2 = 2zw$; and $x = \sqrt{z^2 + \frac{2zw}{m}}$, as before.

Otherwise,
If the lever is supposed to move, the mean

velocity of the end PB will be as $\frac{x}{2}$; that of AP as $\frac{z}{2}$; and that of w as z; and the quantities of matter are w, mz, and mx, which multiplied by their velocities, gives $wz + \frac{mz^2}{2} = \frac{mx^2}{2}$, from which x is found $= \sqrt{\frac{2wz}{m} + z^2}$, as before.

From which we have the following Theorems.

Theo. 2. $z = \sqrt{x^2 + \frac{w^2}{m}} - \frac{w}{m}$.

Theo. 3. $w = \frac{mx^2}{2z} - \frac{mz}{2}$.

Theo. 4. $m = \frac{2wz}{x^2 - z^2}$

EXAMPLE.

Let $z = 2$ feet, $w = 24$lb., $m = 3$lb.; required x, the length which will balance the shorter end, and weight w?

Theo. 1. $x = \sqrt{\frac{2wz}{m} + z^2} = \sqrt{\frac{96}{3} + 4} = 6$.

PROBLEM XI.

Given one end of a straight bar or lever, equally thick, and homogeneous, 6 feet; the weight of a foot in length, 4 pounds; and a

weight of 60 pounds suspended from the shorter; required the length of the shorter, when
it, together with the weight, will balance the
longer?

Here, $x = 6$, $w = 60$, $m = 4$, to find z.

(Per Theo. 2.) $z = \sqrt{x^2 + \dfrac{w^2}{m^2}} - \dfrac{w}{m} =$

$\sqrt{36 + \dfrac{3600}{16}} - \dfrac{60}{4} = 1.15$ feet, the distance of
the weight from the prop.

PROBLEM XII.

Given the longer end 10 feet, the shorter 2
feet, the weight at the shorter end 100 pounds;
required the weight of the lever, so that the
whole may be in equilibrio?

$x = 10$ feet, $z = 2$ feet, $w = 100$lb.; required m?

This is solved by Theo. 4, where $m = \dfrac{2wz}{x^2 - z^2}$

$= \dfrac{400}{96} = 4.166$ pounds, the weight of 1 foot,
which multiplied by 12, the whole length, gives
50 pounds for the whole weight of the lever.

PROBLEM XIII.

FIG. VI.

If a power P, at the distance AC, raises a

weight w, at the distance DC, through DG, in a
given time; what weight M must be placed at
the distance BC, so that P may descend equally
fast, or that the angular velocity may be the
same ?

Let CB $=$ CA $= a$; CD $= b$.

According to the laws of motion, a given
power in a given time moves w through DG;
but when a weight $\frac{wb}{a}$ is applied at B, it is acted
upon by a power (from the property of the lever)
which is to that acting at D as $b : a$; and as $a :$
$b :: w : \frac{bw}{a}$, hence the power and weight have
still the same ratio; of consequence the power
would urge the weight through the same space
in the same time; but the space ought to be
directly as the distance when P moves equally
fast, or when the angular velocity is the same;
and the power being the same, the mass must
be increased in the same inverse ratio as the ve-
locity is diminished, or as $b : a :: \frac{bw}{a} : \frac{b^2w}{a^2} =$
the weight which applied at B, would be raised
through BE by the power P, while w would
be raised through DG by the same power; from
which is derived the third solution to the third
Problem.

Otherwise,

Let v = velocity of the point D.

Then, As $b : a :: v : \frac{va}{b}$ = the velocity of the point B ; but the power is able to communicate to $\frac{bw}{a}$, when placed at B, a velocity v. Therefore the effect produced is $\frac{vbw}{a}$ = M $\times \frac{va}{b}$; or Mva^2 = vwb^2, and M = $\frac{wb^2}{a^2}$, as before.

The same solution is obtained by a different process, in Prob. XIII. part first, where D and d = the distances = a and b ; and there As D^2: d^2 :: F to f: and if F = w, we have $a^2 : b^2$ as w : $\frac{b^2w}{a^2}$.

PROBLEM XIV.

FIG. VII.

Given the weight of a wheel $a\ b$, condensed in the circumference; and a weight p suspended by a cord coiled round the wheel; to determine the velocity of the weight, and of consequence of the circumference of the wheel?

Put w = the weight of the wheel = 20.

$p = 1$. Then, it is evident, the mass to be moved will be $w + p$ = 21 ; and as the wheel is at liberty to rest in any position, the moving power is only p.

Therefore, as $w+p : p$ so is the whole force of gravity to that part which is exerted in moving the wheel: viz. As $w+p : p :: 1 : \frac{p}{w + p}$; or As $21 : 1 :: 1 : \frac{1}{21}$, which multiplied by 16, gives the feet passed over in the first second $=\frac{16}{21} = .7619$ feet, or 9.14 inches.

EXAMPLE II.

Let $p = 2$; then $w + p = 22$, and the accelerating force will be $\frac{2}{22} = \frac{1}{11}$ or 1.4545 feet the first second.

EXAMPLE III.

If $p = 3$; then $w + p = 23$, and $\frac{p}{w+p} = \frac{3}{23}$ $\times$ 16 $= 2.087$ feet.

EXAMPLE IV.

If $p = 4$; then $\frac{p}{w+p} = \frac{4}{24}$, which multiplied by 16, gives 2.66 feet.

In these examples, the mass to be moved is increased by increasing the power, as it makes part of the mass: but in computing central forces, as in the first part of this work, the moving power or force is considered as distinct from the mass moved, which, according to this

Problem, may be illustrated by supposing the mass to diminish as the power increases.

EXAMPLE V.

Let the weight of the wheel be 31, and the power 1. Then $\frac{p}{w+p} = \frac{1}{32}$, which multiplied by 16, gives $\frac{16}{32} = \frac{1}{2}$ foot for the space passed over in the first second.

EXAMPLE VI.

Let the weight be 30 ; the power 2. Then $\frac{p}{w+p} = \frac{2}{32} = \frac{1}{16}$, or 1 foot in the first second.

EXAMPLE VII.

Make the weight 29; the power 3. And we have $\frac{p}{w+p} = \frac{3}{32} = 1.5$ feet the first second.

EXAMPLE VIII.

Let the weight be 28 ; the power 4. And $\frac{p}{w+p} = \frac{4}{32} = \frac{1}{8} = 2$ feet per second.

EXAMPLE IX.

In the four last Examples, required the time in which the power will descend through 20 feet ?

In falling bodies, as the space is to the square of the time, so is any other space to the square of the time in which it is described.

In Ex. 5. As $\frac{1}{2}$ foot : $1''$ :: 20 feet : $\overline{6.32}|''$, which is the time required to fall through 20 feet.

In Ex. 6. As 1 foot : $1''$ :: 20 feet : $\overline{4.47}|''$·

In Ex. 7. As $1\frac{1}{2}$ feet : $1''$ :: 20 feet : $\overline{3.65}|^2$·

In Ex. 8. As 2 feet : $1''$ :: 20 feet : $\overline{3.16}|^2$·

In Examples 5th, 6th, 7th, and 8th, the moving powers are as 1, 2, 3, and 4, and the spaces passed over in a given time are as 1, 2, 3, and 4; but the times in which a given space is described are inversely as the square roots of the impelling powers, viz. in this case, as 1, 1.414, 1.73, and 2, inversely, which numbers are the square roots of the impelling powers. And it may be observed, that when the velocity of a revolving body is said to be as the square root of the impelling power, that power is supposed to descend through a given space, which is always the case with water applied to turn a wheel.

PROBLEM XV.

FIG. VIII.

Given the power, and distance from the axis at which it is applied, and also the diameter and weight of the wheel, to determine the velocity of the power, and of the circumference of the wheel?

Let the distance at which the power acts be $=d$; the radius of the wheel $=r$; its weight $=w$.

Then, from the property of the lever, $\frac{dp}{r}=$ the force with which p acts upon the wheel, or it expresses the moving power. The mass moved is $w+\frac{d^2p}{r^2}$ by which divide $\frac{dp}{r}$ and we have $\frac{dpr}{r^2w+d^2p}$ for the accelerating force at the circumference of the wheel.

And as $r:d::\frac{dpr}{r^2w+d^2p}:\frac{d^2pr}{r^3w+d^2pr}$, which divided by r, gives $\frac{d^2p}{r^2w+d^2p}$ for the velocity of p.

EXAMPLE I.

Let $p=4$; $d=3$; $r=9$; $w=48$.

Then, $\frac{dpr}{r^2w+d^2p}=\frac{3\times4\times9}{81\times48+9\times4}=\frac{108}{3924}=\frac{3}{109}$ of gravity, which multiplied by 16, gives $\frac{48}{109}$ of a foot the first second.

And as $9:3::\frac{48}{109}:\frac{16}{109}$, the space passed over by p in the same time.

c

EXAMPLE II.

Again, let $r = 18$, the rest as before. And we have $\frac{216}{15588} = \frac{6}{433}$ for the velocity of the wheel, which multiplied by 16, gives the feet passed over in the first second.

In this Example, the power p, reduced to the wheel, does not afford so much resistance as in the first; for there $\frac{d^2p}{r^2} = \frac{4}{9}$; but here $\frac{d^2p}{r^2} = \frac{1}{9}$; hence if w, in this Example, had been increased by $\frac{3}{9}$, the accelerating force would have been exactly half of what it is found in the first Example: but without that addition it differs $\frac{1}{8266}$, or is so much above one-half of the other.

EXAMPLE III.

Let the distance $d = 1$; the radius of the wheel $r = 84$; the weight $w = 9$; $p = 8$; required the time in which p descends through 24 feet?

$\frac{dpr}{r^2w + d^2p} = \frac{84}{7939}$, which multiplied by 16, gives .16929 feet the first second.

And as $r : d :: .16929 : .002015$, the feet passed over by p in the same time.

And as $.002015$ feet : $1''$ :: 24 feet : $\overline{109.1}]^{2''}$, the time required.

On the Centre of Gyration.

PROBLEM XVI.

FIG. IX.

Given a straight bar AB, equally thick, &c. to find o, the centre of gyration, which if struck will communicate the same angular velocity to the bar as if the whole bar was collected in that point?

The force of any particle revolving round a centre is as that particle multiplied by the square of its velocity, or of its distance from the centre of motion; of consequence, the force required to destroy that motion must be equal to it.

Put AC $= z$; BC $= x$; CO $= y$, the centre of gyration sought.

Then z multiplied by the square of its distance $\dot{z}^2 = z^2 \dot{z}$, and its fluent $\frac{z^3}{3} =$ the force of the end AC, and $\frac{x^3}{3}$ will be the force of the end BC; but when x and z are removed to o, their force will be $\overline{x + z} \times y^2$; which force must be equal to $\frac{x^3}{3} + \frac{z^3}{3}$: viz. $\overline{x+z} \times y^2 = \frac{x^3 + z^3}{3}$; and

$$y = \sqrt{\frac{x^3 + z^3}{3x + 3z}}.$$

EXAMPLE.

Let $z = 7$; $x = 14$. Then, $y^2 = \frac{2744+343}{42+21}$ $= \frac{3087}{63}$, and $y = \sqrt{49} = 7$, the distance of the centre of gyration from the centre of motion.

PROBLEM XVII.

FIG. X.

Given three bodies A, B, and D, connected by a line supposed without weight, to find the distance of N the centre of gyration, from C the centre of motion?

Let $CA = a$; $CB = b$; $CD = d$; $CN = y$.

$Aa^2 + Bb^2 + Dd^2 = \overline{A + B + D} \times y^2$; for $\frac{Aa^2}{y^2} + \frac{Bb^2}{y^2} + \frac{Dd^2}{y^2} =$ the sum of all the forces reduced to N, which, by the Problem, must be equal to all the bodies placed at N and multiplied by $\overline{CN^2}$; hence $y = \sqrt{\frac{Aa^2 + Bb^2 + Dd^2}{A+B+D}}$.

EXAMPLE.

Let $A = 2$; $a = 10$: $B = 4$; $b = 6$; $D = 6$; $d = 4$.

Then, $\sqrt{\frac{200+144+96}{2+4+6}} = 6.05 = y$, the distance sought.

PROBLEM XVIII.

FIG. XI.

Given the length and weight of a bar, two weights A and B placed at the ends, and c the centre of motion, to find N the centre of gyration?

Let $b =$ the weight of AC; $a =$ the weight of BC; $x =$ AC; $z =$ BC; $w =$ the weight at A; $p =$ the weight at B.

Then, $wx^2 + \dfrac{bx^2}{3} + pz^2 + \dfrac{az^2}{3} =$ the force of the whole revolving round c, to which $\overline{w + p + a + b} \times \overline{\text{CN}}^2$ must be equal; hence, CN, or $y, = \sqrt{\dfrac{3wx^2 + bx^2 + 3pz^2 + az^2}{3 \times \overline{w+p+a+b}}}$.

EXAMPLE.

Let BC $= a = 1$; AC $= b = 2$; the weight at A $= w = 1$; the weight at B $= p = 1$; $x = 14$; $z = 7$.

And we have $y = \sqrt{\dfrac{588 + 392 + 147 + 49}{3 \times \overline{1+1+1+2}}} = \sqrt{\dfrac{1176}{15}} = 8.85$, at which distance, if the whole mass in the system was collected, it would require the same force to give it the same angular motion, or to destroy the motion, as if the bodies had been fixed as in the Problem.

On the Centre of Gravity.

This centre, in any body or system of bodies, if sustained, the whole remains at rest. This has been treated on by so many authors, and is in general so well understood, that a few observations will be sufficient.

PROBLEM XIX.

FIG. XII.

Given two bodies A and B, connected by the line AB, to find C, their common centre of gravity?

Let $AB = a$; $AC = x$.

Then, $A \times x = B \times \overline{a-x}$, or $Ax = Ba - Bx$; and $Ax + Bx = Ba$: hence, $x = \frac{Ba}{A+B}$.

EXAMPLE.

Let $AB = a = 100$; $A = 40$; $B = 10$.

Then, $\frac{100 \times 10}{40 + 10} = 20 = x$, the distance of the centre of gravity from the larger body.

PROBLEM XX.

Given the length of an homogeneous beam, the weight of a foot in length, and a weight suspended from one end, to find the centre of gravity ?

Let n = the length in feet.
m = the weight of a foot in length.
w = the weight suspended.
x = the longer end.
z = the shorter end $= n - x$.

Then $m\dot{z}z + w\dot{z}$ = the momentum of the shorter end; and $m\dot{x}x$ = the momentum of the longer end. And by making their fluents equal, according to the Problem, we have $\frac{mx^2}{2} = \frac{mz^2}{2} + wz$, and by substituting $n-x$ for z, we have $\frac{mx^2}{2} = \frac{mn^2-2mnx+mx^2}{2} + wn - wx$, from which x is found $= \frac{mn+2w}{mn+w} \times \frac{n}{2}$.

EXAMPLE.

Let $n = 20$; $m = 100$; $w = 1000$.

Then we have $x = \frac{2000+2000}{2000+1000} \times \frac{20}{2} = 13\frac{3}{9}$, or $13\frac{1}{3}$.

PROBLEM XXI.

FIG. XIII.

Given the length of a beam AB $= 2n$; the weight of a foot in length $= m$; the weight w, and its distance from A $= a$: the weight P, and its distance from B $= b$, to find G, the centre of gravity of the whole?

Let $x =$ AG; $z =$ BG $= 2n - x$.

Then the momentum of w will be $=$ w $\times \overline{x - a}$, the momentum of the end x will be $= \frac{mx^2}{2}$, the momentum of P $= p \times \overline{z - b}$, and that of z, or its equal $2n - x$, $= \frac{4mn^2 - 4mnx + mx^2}{2}$; hence we have $mx^2 + 2wx - 2wa = 4mn^2 - 4mnx + mx^2 + 4pn - 2px - 2bp$; from which we find $x = \frac{2mn^2 + 2pn + wa - bp}{2mn + w + p}$

If the weights w and P are at the ends of the beam, then will x and $2n - x =$ their distances from the centre of gravity, and the theorem will be $x = \frac{4mn^2 + 4pn}{4mn + 2w + 2p} = \frac{mn + p}{2mn + w + p} \times 2n$.

EXAMPLE.

$n = 10$; $m = 4$; $w = 10$; $p = 3$; $a = 4$; $b = 1$. Then $x = \frac{1600 + 120 + 80 - 6}{160 + 20 + 6} = 9.6451$, the distance of the centre of gravity from the end A.

Of the Centre of Percussion.

As in bodies at rest the whole weight may be considered as collected in the centre of gravity, so in bodies in motion the whole force may be considered as concentrated in the centre of percussion, or centre of force.

PROBLEM XXII.

FIG. XIV.

In a straight rod O A, required C, the centre of percussion?

Put $a =$ O A, the length of the rod; $x =$ O C, the distance of the centre of percussion from the point of suspension; O G $=$ the distance of the centre of gravity $= \frac{a}{2}$. And the weight of the rod multiplied by the distance of the centre of gravity from the point O, will be equal to the force of the rod divided by the distance of the centre of percussion from O.

Therefore, $\frac{a^2}{2} = \frac{a^3}{3x}$, and $3a^2x = 2a^3$, which divide by a^2, and we have $3x = 2a$, or $x = \frac{2a}{3}$.

And as the stroke is the same as if the whole weight of the rod was condensed in this point,

H

it is also the centre of oscillation ; for, if it be considered as a pendulum, the whole force of the rod is exerted in the point c, to move the pendulum, and of consequence the time of vibration must be the same as if the whole weight was actually collected in that point.

PROBLEM XXIII.

FIG. XV.

Let the centre of motion, or point of suspension, be at a distance from the end of the rod, to find the centre of percussion?

Let $OB = a = 8$; $OA = b = 4$; $OC = x$.

Then, $\frac{a-b}{2} = OG$, the distance of the centre of gravity from o, which multiplied by the weight $a+b$, gives $\frac{a^2-b^2}{2} = \frac{a^3+b^3}{3x}$; and $3x \times \overline{a^2-b^2} = 2 \times \overline{a^3+b^3}$, or $x = \frac{a^3+b^3}{a^2-b^2} \times \frac{2}{3} = 8$; hence, in this Example, c coincides with B.

PROBLEM XXIV.

FIG. XVI.

Given the length and weight of a rod, and the weight of a ball at the end, to find the centre of percussion, or oscillation?

Put $n =$ the length of the rod; $m =$ the weight of a foot; $w =$ the weight of the ball; $y = $ oo, the distance of the centre of gravity from o; $x = $ oc, the distance of the centre of oscillation.

Then, $y = \frac{mn + 2w}{mn + w} \times \frac{n}{2}$; (see Prob. XX.) which multiplied by the weight $mn + w$, gives $\frac{mn^2 + 2wn}{2} = \frac{mn^3 + 3wn^2}{3x}$; and $x = \frac{mn^3 + 3wn^2}{mn^2 + 2wn} \times \frac{2}{3}$.

Otherwise,

The weight of the rod is mn, which multiplied by half its length, gives $\frac{mn^2}{2}$ for the moment of the rod; and $n \times w$ is the moment of the ball. Also, mn, the weight of the rod, multiplied by the square of the length, n^2, gives mn^3, which divided by 3, gives the whole force of the rod. And $w \times n^2$ is the force of the ball; and the sum of their forces divided by the sum of their moments, gives the distance of the centre of percussion from the point of suspension, $= \frac{\frac{1}{3}mn^3 + wn^2}{\frac{1}{2}mn^2 + wn}$, the same as above.

PROBLEM XXV.

FIG. XVII.

Given the length and weight of a rod, and

the weights and places of two balls fixed to it; required the centre of percussion, or oscillation?

Put $m=$ the weight of a foot; $n=$ the length ow; $a = $ op; $x = $ oc.

Then $\frac{mn^2}{2}=$ the momentum of the rod; that of the weight p is ap; and that of w $= n$w: the sum of the whole is $\frac{mn^2}{2} + a$p $+ n$w.

The force of the rod is $\frac{mn^3}{3}$; the force of the weight p is p $\times a^2$; the force of the weight w is w $\times n^2$; and the sum is $\frac{mn^3}{3} + $ p$a^2 + $ wn^2, which divided by the sum of the momentum, gives $x=\frac{mn^3 + 3\text{p}a^2 + 3\text{w}n^2}{mn^2 + 2\text{p}a + 2\text{w}n} \times \frac{2}{3}$, the same as if $mn + $ w $+ $ p, the whole weight of the rod and balls, were multiplied by $\frac{mn^2 + 2n\text{w} + 2a\text{p}}{2mn + 2\text{w} + 2\text{p}}$, the distance of the centre of gravity from the point of suspension.

Note.—If the distance of the centre of oscillation from the centre of the system, or point of suspension, be multiplied by the distance of the centre of gravity from the same point, the square root of the product will be the distance of the centre of gyration: for instance,

$$\frac{mn^3 + 3pa^2 + 3wn^2}{mn^2 + 2pa + 2wn} \times \frac{2}{3} \times \frac{mn^2 + 2pa + 2nw}{2mn + 2p + 2w} =$$

$$\frac{mn^3 + 3pa^2 + 3wn^2}{2mn + 2p + 2w} \times \frac{2}{3} = \frac{mn^3 + 3pa^2 + 3wn^2}{3 + w + p + mn},$$

which last expression is the square of the centre of gyration, as found by Problem XVIII.

EXAMPLE.

Let $n = 12 = $ ow, w $= 2$; $a = 6 = $ op, p $= 2$; $m = 1$.

Then $\frac{mn^2 + 2pa + 2wn}{2mn + 2p + 2w} = \frac{216}{32} = 6.75$, the distance of the centre of gravity from the point o.

And $\frac{mn^3 + 3pa^2 + 3wn^2}{mn^2 + 2pa + 2wn} \times \frac{2}{3} = \frac{5616}{648} = 8\frac{2}{3}$, the distance of the centre of oscillation or percussion.

Also, $\frac{mn^3 + 3pa^2 + 3wn^2}{3 \times w + p + mn} = \frac{2808}{48} = 58.5$, the square root of which is 7.6485, the centre of gyration.

Put c $= 6\frac{3}{4}$, the centre of gravity.
p $= 8\frac{2}{3}$, the centre of percussion.
g $= 7.6485$, the centre of gyration.

Then GP $=$ cc, and As g : c :: c : p $= \frac{c^2}{g}$
$= \frac{58.5}{6.75} = 8\frac{2}{3}$; and p : c :: c : g $= \frac{c^2}{p} = \frac{58.5}{8\frac{2}{3}}$
$= \frac{351}{52} = 6\frac{3}{4}$.

PROBLEM XXVI.

FIG. XVIII.

Given the length and weight of a rod, the centre of suspension s, a weight w at the bottom, and another p at the top, to find the centre of percussion?

Let m = the weight of a foot in length : n = sw feet ; a = sp feet.

Then, the whole force of the rod and balls will be $\frac{mn^3 + 3wn^2 + ma^3 + 3pa^2}{3}$: the momentum of the end sw will be $\frac{mn^2}{2} + wn$: the momentum of the end ps will be $\frac{ma^2}{2} + ap$. And by dividing the sum of their forces by the difference of their momentums, we have $\frac{mn^3 + 3wn^2 + ma^3 + 3pa^2}{mn^2 + 2wn - ma^2 - 2pa}$ $\times \frac{2}{3}$ = the centre of percussion sought.

EXAMPLE.

Let $m = 16$; w = p = 39 dwts.; $a = \frac{1}{3}$ foot; $n = \frac{2}{3}$ foot. Then, from the above Theorem,
$$\frac{\frac{128}{27} + \frac{468}{9} + \frac{16}{27} + \frac{117}{9}}{\frac{64}{9} + \frac{156}{3} - \frac{16}{9} - \frac{78}{3}} \times \frac{2}{3} = \frac{1899}{846} \times \frac{2}{3} = 1.4964$$
feet, the distance of the centre of percussion or oscillation from the point of suspension. c, the centre of percussion, is, in this case, fur-

ther removed from the point of suspension
than the weight w.

Note.—If s is the centre of the rod, or if
sb = sw, and a = the distance of p, the expression for the centre of oscillation will be a
little shorter, viz. $= \dfrac{2mn^3 + 3wn^2 + 3pa^2}{2wn - 2pa} \times \dfrac{2}{3}.$

Let every thing be the same as in the last
Example, only sb = sw = n. Then, from
the last Theorem, the distance of the centre of
oscillation from s is found = 22.917 inches.

Note 2.—If the weight of the rod is not con-
sidered, the Theorem becomes $\dfrac{wn^2 + pa^2}{wn - pa} = sc,$
the distance of the centre of oscillation from s.

Note 3.—If m = the weight of an inch,
and the distances are given in inches, we have
the answer in inches.

PROBLEM XXVII.

Given a straight bar, equally heavy, to determine the point of suspension, when the vibrations are performed in the least time possible;
or when the distance of the centre of oscillation from the axes of motion is a minimum?

Let n = the whole length of the bar; x =
the length of the shorter end.

Then will $\frac{n^3-3n^2x+3nx^2}{n^2-2nx} \times \frac{2}{3}$ express the distance of the centre of oscillation from the point of suspension, which is to be a minimum.

Its fluxion is $6nxx-3n^2x \times \overline{n^2-2nx}-2nx$ $\times n^3-3n^2x+3nx^2 = 0$. And $6nx-3n^2-$ $12x^2+6nx = 6nx-2n^2-6x^2$; from which we get $6nx-6x^2 = n^2$; and by completing the square, we have $x^2-nx+\frac{n^2}{4} = \frac{n^2}{4}-\frac{n^2}{6}$: and $x = \frac{n}{2}-n\sqrt{\frac{1}{12}}$.

EXAMPLE.

Let the length of the rod be 40 inches $= n$. Then $\frac{40}{2}-40\sqrt{\frac{1}{12}} = 8.4532 = x$, the distance of the centre of suspension from the end, when it vibrates in the least time possible.

The centre of oscillation will be found $=$ 23.0926 inches, equally distant from the lower end of the rod as the centre of suspension is from the top thereof. Hence, if the centre of oscillation is made the centre of motion, the centre of motion becomes the centre of oscillation, and the time of vibration remains the same.

In a straight rod, the centre of oscillation is two-thirds of its length from the top, when it is suspended at the top; but if the bar is suspended at one-third of its length from the end, the centre of oscillation is at the bottom, or lower end, and the vibrations are performed in the same time. The centre found in this Problem would be the centre of gyration, if the centre of gravity were the centre of suspension. In this case, the centre of gravity is in the middle of the rod ; and the centre of gyration is $n\sqrt{\frac{1}{12}}$, which taken from $\frac{n}{2}$ gives the centre of suspension, as before.

When a pendulum, or vibrating body, is placed in a horizontal position, the initial velocity of the centre of oscillation is the same as that produced by the action of gravity on bodies left to fall in straight lines towards the earth.

Let AB (FIG. XIX.) be a pendulum, s the centre of motion, B the centre of oscillation, then the end B will begin to descend equally fast as a body unconnected with the pendulum. For, if the end AS affords resistance, it throws the centre of oscillation to B, which if AS were removed would be at c. The point c would, in that case, begin to move with the velocity of a falling body, and B would move faster, as it is further from the centre of motion.   Or if the

I

centre of oscillation were removed beyond the
end of the vibrating body, (see Fig. XVIII.) the
weight w would descend with a less velocity,
or it would be As sc : sw :: 16 : 16 × $\frac{sw}{sc}$ =
the initial velocity of w. Hence, in any system, having found the centres of gravity and
oscillation, we find their tendency to motion,
or the initial accelerating force.

PROBLEM XXVIII.

Given the length of a beam, 20 feet; its
weight, 200lb ; a weight, w, = 60lb. at one
end; and a power P, = 100, at the other end;
it is required to assign the velocity of P, which
is equal to that of w when the beam, equally
thick, is suspended by the centre of gravity?

A given power applied to any part of the
beam will communicate the same velocity as if
the whole mass was collected in the centre of
gyration.

Let a = 10 feet, half the length of the beam.
    m = 100lb. half the weight of the beam.

The centre of gyration in the beam itself,
without the weight w and P, is $\sqrt{\frac{2a^3}{ca}} = \sqrt{\frac{a^2}{3}}$
$= a\sqrt{\frac{1}{3}}$. (See Prob. XVI.)

$\sqrt{\frac{1}{3}}$ = .57735, which × 10 = 5.7735, the distance of the centre of gyration from the centre of the beam.

Therefore, whether the whole mass is dispersed equally through a beam 20 feet long, whether it is collected in the centre of gyration, or whether it is diffused through a circle, the radius of which is 5.7735 feet—yet a given power, applied at a given distance, would, in the same time, communicate the same angular velocity, or would turn it equally fast round its axis.

Next, the resistance of the beam must be reduced to the extremity of the arm, where the power acts, (see Prob. XIII.) which is done by multiplying the weight by the square of the distance of the centre of gyration, and dividing by the square of the distance of the power P from the centre of the beam. Thus the square of the distance of the centre of gyration is $\frac{a^2}{3}$, and the weight is $2m$; their product is $\frac{2ma^2}{3}$, which divided by a^2, gives $\frac{2m}{3}$ for the whole resistance of the beam reduced to the end. And, as $2m =$ the whole weight, the resistance which it affords to a power acting at the end is $= \frac{1}{3}$ of its weight.

We have therefore $w + p + \frac{2m}{3} =$ the mass to be moved; and $100 - 60 = 40 =$ the moving power. And the space passed over in one second, compared with gravity, is $\frac{p - w}{p + w + \frac{2m}{3}}$ $= \frac{100 - 60}{100 + 60 + 66.6} = .17647$, which multiplied by 16, gives 2.822 feet the first second.

Otherwise,

If we take 40, the moving power, from 100, we shall have each end loaded with 60lb. The centre of gyration of the beam thus loaded will be expressed by $\overline{\frac{2wa^2 + \frac{2ma^2}{3}}{2w + 2m}}\Big|^{\frac{1}{2}} = \sqrt{58.33}$: the mass is $2w + 2m$, which multiplied by $\frac{2wa^2 + 2ma^2}{3}$ $\frac{2wa^2 + \frac{2ma^2}{3}}{2w + 2m}$ gives $2wa^2 + \frac{2ma^2}{3}$, which divided by a^2, the distance of the power, quotes $2w + \frac{2m}{3}$ for the whole resistance reduced to the end of the beam, to which add $p - w$, and we have the whole mass to be moved; by which divide the moving power, $p - w$, the quotient is the accelerating force, $= \frac{p - w}{2w + \frac{2m}{3} + p - w} = \frac{p - w}{\frac{2m}{3} + p + w} = \frac{40}{120 + 66\frac{2}{3} + 40} = \frac{40}{226\frac{2}{3}} = .17647$, as before.

Otherwise,

If we divide the distance of the weight w by the distance of the centre of oscillation, we have the initial velocity.

If $2m =$ the whole weight of the beam; and $2a =$ the length, as before;

Then will $\dfrac{2am + 3aw + 3aP}{3P - 3w}$ express the distance of the centre of oscillation, by which divide a, the distance of w, and we have $\dfrac{3P - 3w}{2m + 3w + 3P} =$ 17647, the accelerating force compared with gravity, the same as before.

PROBLEM XXIX.

Given one end of a lever 6 feet, the other 10, the weight of the whole $= 160$lb. a weight w $= 300$lb. at the shorter end, and a power P $= 500$lb. at the longer end; required the accelerating force?

Let $a = 6$, the shorter end.
  $b = 10$, the longer end.
  $m = 160$, the whole weight.

Then, the square of the distance of the centre of gyration (Prob. XVI.) is $\dfrac{a^3 + b^3}{3a + 3b}$; which multiplied by the weight m, and divided by the square of the distance of P, gives $\dfrac{a^3 + b^3}{3a + 3b} \times \dfrac{m}{b^2} =$

the whole resistance of the lever at P.   To this add the weight w, reduced to P, and the power P ; and we have the whole resistance at P $=$ $\frac{a^3 + b^2}{3a + 3b} \times \frac{m}{b^2} + \frac{a^2 w}{b^2} + $ P $= 648.533.$

Then, as 16, the whole length of the lever, is to 160, the whole weight, so is 10, the longer arm, to 100, its weight, which taken from 160, leaves 60, the weight of the shorter.

Then (Prob. XXI.) $\frac{2wa + 60a - 100b}{2b} = 148,$ which, if placed at P, would exactly balance the weight w ; this, taken from P, leaves the moving power $= 352,$ which divide by the whole mass, and we have 648.533) 352 (.54276, which multiplied by 16, gives 8.684, the feet through which P would descend in one second, if the direction was perpendicular.

Otherwise,

Let $m =$ the weight of a foot in length.

Then, $\frac{mb^3 + 3\text{P}b^2 + ma^3 + 3wa^2}{mb^2 + 2\text{P}b - ma^2 - 2wa} \times \frac{2}{3} =$ the centre of oscillation $= \frac{608}{33}$; by which divide 10, the distance of P, and we have $\frac{330}{608} = 54276$; and as b is to $a,$ so is the space described by P to that described by w.

PROBLEM XXX.

Given, $a=6$ feet, the shorter end of a lever; $b=10$ feet, the longer end; $m =$ the weight of the longer end, $= 100$lb.; the weight of the shorter $=n=60$; $w=78$; and let the power P be suspended from the longer arm at a distance $d = 7$: required the accelerating force at P?

The centre of oscillation (see Prob. XXV.) will be $= \frac{2mb^2 + 6Pd^2 + 2Pa^2 + 6a^2w}{3mb + 6Pd - 3na - 6aw}$; by which divide d, and the quotient is $\frac{3mbd + 6Pd^2 - 3nad - 6awd}{2mb^2 + 6Pd^2 + 2Pa^2 + 6a^2w}$ $=$ the accelerating force at P; $= .359$ parts of gravity. Or, by the process in Prob. XXVII. if $m =$ the weight of a foot, then $\frac{2wa + ma^2 - mb^2}{2d}$ $= 211\frac{6}{14}$, the weight which, placed at P, would balance w: hence, the accelerating force, considered alone, is $500 - 211\frac{6}{14} = 288\frac{8}{14}$. The mass to be moved, reduced to P, $= \frac{a^3 + b^3}{3a + 3b} \times \frac{m}{d^2}$ $+ \frac{a^2w}{d^2} + P = 803.129$, by which divide $288\frac{8}{14}$, and we have $.359$, as before.

PROBLEM XXXI.

Given, the radius of a solid wheel, equally thick, $= d = 10$ inches; the radius of the axis $= b = 1$ inch; a power P $= 2$lb. suspended

from the wheel; and a weight w $=$ 16lb. from
the axis; the weight of the wheel $=$ 4lb. $= m$:
required the space descended through by the
power compared with falling bodies?

First, to find the centre of gyration in the
wheel;

Let $x =$ the distance of a particle from the
  centre.

$q = .7854.$

Then will $4q \times 2x =$ the circumference of a
circle or ring, the diameter of which is $2x$.

The fluxion of its area will be $8qxx$. And,
As $8qxx : 4qd^2$, the whole area, . : the weight
of any particle a : the whole weight of the
wheel, $= \dfrac{4aqd^2}{8qxx} = m$; and if a, or its equal,
$\dfrac{8mqxx}{4qd^2}$, be multiplied by x^2, the square of the
distance of the particle, it gives $\dfrac{8mqx^3x}{4qd^2} = \dfrac{2mx^4}{4d^2}$
(and when $x = d$) $= \dfrac{md^2}{2} = y^2m$; and $y^2 = \dfrac{d^2}{2}$
or $y = d\sqrt{\tfrac{1}{2}} =$ the distance of the centre of
gyration from the centre of the wheel $= 10 \times$
70716 $= 7.0716$ inches.

The weight of the wheel multiplied by the
square of the distance of the centre of gyration,

$(= \frac{d^2}{2} \times m)$ and divided by the square of the radius, gives the resistance which it affords at the circumference $= \frac{m}{2}$. The weight w, reduced to the circumference, is $\frac{wb^2}{d^2}$. Suppose the weight of the axis to be 5lb. $= g$, the resistance at its surface will be $\frac{b^2}{2} \times g$, and at the circumference of the wheel $= \frac{gb^2}{2} \times \frac{b^2}{d^2}$, these added, give $\frac{gb^4}{2d^2}$ $+ \frac{m}{2} + \frac{wb^2}{d^2} + $P, the whole resistance; by which divide P $- \frac{bw}{d}$, the moving power, (see Prob.

XV.) and we have $\dfrac{P - \dfrac{bw}{d}}{\dfrac{gb^4}{2d^2} + \dfrac{m}{2} + \dfrac{wb^2}{d^2} + P} = \dfrac{4}{4.1625}$ $= .096$, which multiplied by 16, gives 1.536 feet the first second.

If it is required to raise the weight w through a given space in the least possible time, by the action of the power P, it will be necessary to determine the diameter of the wheel to that of the axle. Let $b =$ the diameter of the axle, and z that of the wheel, which is variable. The accelerating force will be expressed by $\frac{2pz^2 - 2bwz}{2wb^2 + mz^2 + 2pz^2}$, $=$ the space through which P descends in 1 second, compared with gravity, (1). And as the square root of that space is to

K

1 second, the time in which it is described, so is the square root of any other space s to $\sqrt{s} \times$ $\sqrt{\dfrac{2wb^2 + mz^2 + 2pz^2}{2pz^2 - 2bwz}}$, the time of describing it, which time is to be a minimum, or the least possible: its fluxion is $\overline{2mz\dot{z} + 4pz\dot{z}} \times$ $\overline{2pz^2 - 2bwz} - \overline{4pz\dot{z} - 2bw\dot{z}} \times \overline{2wb^2 + mz^2}$ $+ 2pz^2 = 0$. From which we get $mz^2 + 2pz^2$ $- 4pbz = 2wb^2$; and $z^2 - \dfrac{4pb}{2p + m} \times z =$ $\dfrac{2wb^2}{2p + m}$: if $\dfrac{4pb}{2p+m} = n$, then $z = \sqrt{\dfrac{2wb^2}{2p + m} + \dfrac{n^2}{4}}$ $+ \dfrac{n}{2} =$ the radius of the wheel required. If the weight of the wheel is not considered, the Theorem becomes $z = b\sqrt{\dfrac{w^2}{p^2} + \dfrac{w}{p}} + \dfrac{bw}{p}$.

If the power P is suspended from the axle, and the weight from the wheel; and if the radius of the axle be 1, that of the wheel z, the weight of the wheel m, and its centre of gyration y; then will $\dfrac{P - wz}{y^2m + wz^2 + P} \times 16$ be the space passed over by P in 1 second; which multiplied by z, gives the space passed over by w in the same time. Or if P is suspended from the wheel, and w from the axle, then the space passed over by P will be $\dfrac{Pz^2 - wz}{y^2m + w + Pz^2} \times 16$; which divided by z, gives the velocity of w.

*On the Maximum of Machines, &c. at the end of
a given time.*

PROBLEM XXXII.

Given the weight of a beam 200lb. the arms
of equal lengths, a weight of 100lb. suspended
from one end; it is required to ascertain the
weight suspended from the other, so that when
it is multiplied by its velocity the product shall
be a maximum?

Put $m = 200$lb. the weight of the beam.

$\quad p = 100$lb.

$\quad x =$ the weight sought.

Then will $\dfrac{p-x}{\dfrac{m+p+x}{3}}$ (see Prob. XXVII.) be

the accelerating force; and $\dfrac{px-x^2}{\dfrac{m+p+x}{3}} =$ the mo-

mentum of x, which in fluxions is $\overline{p\dot{x} - 2x\dot{x}} \times$

$\overline{\dfrac{m}{3} + p + x - x} \times \overline{p\dot{x} - x^2} = 0$; from which

we get $x^2 + 2px + \dfrac{2mx}{3} = \dfrac{pm}{3} + p^2$; by putting

the coefficients of x, $2p + \dfrac{2m}{3} = n$, and com-

pleting the square, we find $x = \sqrt{\dfrac{pm}{3} + p^2 + \dfrac{n^2}{4}}$

$- \dfrac{n}{2} = 44.152.$

$\dfrac{p-x}{\dfrac{m+p+x}{3}}$, the space passed over in the first

second, $=.263$, which multiplied by the weight x, gives the product 11.7 ; which product is greater than can be obtained by assuming x either more or less than given by the Theorem. But if the weight of the beam is altered, although the power p remains the same, the weight x will be found to vary. For example, if the weight of the beam is 30, then x (from the above Theorem) will be found $= 42$, and its velocity $.381$, which $\times 42$, gives 16, the momentum. So that by diminishing the weight of the beam, the same power, in an equal time, produces a greater effect on x; not that the absolute effect is greater, but the additional force which is now exerted on x was before employed to move the beam.

Again, let the weight of the beam be 360, $p = 100$ as before, and we shall find, as above, the value of $x = 45.33$, its velocity $= .206$, and their product, or the momentum, $= 9.337$.

Therefore, when the effect is the greatest, the weight x increases as the beam becomes heavier; but the velocity diminishes in a greater ratio: of consequence, the effect produced by the power p, in raising weights, is diminished by increasing the weight of the machine.

Given the length and weight of a beam, the power suspended from the longer arm, to determine the weight at the end of the shorter, when the effect is a maximum?

Put a = length of the shorter = 6.
  n = its weight = 60.
  b = the longer arm = 10.
  m = its weight = 100.
  p = 500.

Then will $\dfrac{amb + 2apb - na^2 - 2a^2w}{mb^2 + 3pb^2 + na^2 + 3a^2w} \times \dfrac{3}{2}$ = the accelerating force at w. Let the known terms in the numerator be put = N, and those in the denominator = Q, and we shall have $\dfrac{N - 2a^2w}{Q + 3a^2w}$ $\times w$ a maximum; in fluxions $\overline{N\dot{w} - 4a^2w\dot{w}} \times \overline{Q + 3a^2w} - 3a^2\dot{w} \times \overline{Nw - 2a^2w} = 0.$

And $w = \dfrac{1}{a^2}\sqrt{\dfrac{QN}{6} + \dfrac{Q^2}{9}} - \dfrac{Q}{3a^2}.$

$amb + 2apb - na^2 = N = 63840.$

$mb^2 + 3pb^2 + na^2 = Q = 162160.$

$\frac{QN}{6} = 1725382400.$

$\frac{Q^2}{9} = 2921762844.44$

Sum 4647145244.44, the square root of this is 68169.9, which multiplied by $\frac{1}{a^2}$ ($= \frac{1}{36}$) is 1893.61; from which subtract 1501.4814 ($\frac{Q}{3a^2}$) leaves $392.13 = w$, the weight required.

The same conclusion may be brought out by a different process. If $\sqrt{\frac{a^3 + b^3}{3a + 3b}} = r$, the centre of gyration in the beam; then will

$$\frac{P + \frac{m}{2} - \frac{an}{2b} - \frac{aw}{b}}{\frac{m+n}{b^2} \times r^2 + P + \frac{a^2 w}{b^2}} = \text{ the velocity of P.}$$

Then let $P + \frac{m}{2} - \frac{an}{2b} = N$; and $\frac{m + n}{b^2} \times r^2 + P = Q.$

And we have $\dfrac{Nw - \dfrac{aw^2}{b}}{Q + \dfrac{a^2 w}{b^2}} = \text{ the effect; which}$ by the Problem must be the greatest possible.

The fluxion of this expression is $\overline{N\dot{w} - \dfrac{2aw\dot{w}}{b}}$
$\times \overline{Q + \dfrac{a^2 w}{b^2}} - \dfrac{a\dot{w}}{b^2} \times \overline{Nw - \dfrac{aw^3}{b}} = 0.$ Whence
$w = \dfrac{b}{a} \sqrt{\dfrac{NQb}{a} + \dfrac{Q^2 b^2}{a^2}} - \dfrac{Qb^2}{a^2}.$

The value of N and Q in this process will be different from what they are found in the last; but the value of w will be exactly the same.

For example,

$$P + \frac{m}{2} - \frac{an}{2b} = N = 532.$$

$$\frac{m+n}{b_2} \times r^2 + P = Q = 540.533.$$

$$\frac{NQb}{a} = 479272.88.$$

$$\frac{Q^2 b^2}{a^2} = \underline{811600.79.}$$

Sum 1290873.67.

$\sqrt{1290873.67} = 1136.16$; which multiplied by $\frac{b}{a}$, gives 1893.6

Subtract $\underline{1501.4} = \frac{Qb^2}{a}$

Rem.   . 392.2 $= w$, the answer as before.

PROBLEM XXXIV.

Let the weight P be suspended from the arm, at any distance from the end; required the weight of w, in case of a maximum?

Let $a =$ the shorter arm.
$b =$ the longer.
$d =$ the distance of P from the centre.
$n =$ the weight of the shorter end.
$m =$ the weight of the longer.

The centre of oscillation of the whole will be expressed by $\dfrac{2mb^2 + 6\mathrm{P}d^2 + 2na^2 + 6a^2w}{3mb + 6\mathrm{P}d - 3na - 6aw}$, by which divide a, the distance of the weight w, and we have $\dfrac{3amb + 6ard - 3na^2 - 6a^2w}{2mb^2 + 6\mathrm{P}d^2 + 2na^2 + 6a^2w} =$ the accelerating force, or initial velocity of w; which, when multiplied by w, must be a maximum: its fluxion is $\overline{amb\dot{w} + 2apd\dot{w} - na^2\dot{w} - 2a^2w\dot{w}} \times$
$\overline{mb^2 + 3pd^2 + na^2 + 3a^2w} - 3a^2\dot{w} \times$
$\overline{ambw + 2apdw - na^2w - 2a^2w^2} = 0$, which multiplied and reduced, gives $6a^4w^2 + a^4nw + 3na^3w + 12a^2pd^2w + 4ma^2b^2w = am^2b^3 + 2apdmb^2 + 3ambpd^2 + 6ap^2d^3 + a^3mbn + 2a^3pdn - nma^2b^2 - 3na^2pd^2 - n^2a^2$.

Let the coefficients of the first power of w, $\dfrac{n}{6} + \dfrac{n}{2a} + \dfrac{2pd^2}{a^2} + \dfrac{2mb^2}{3a^2}$, $= c$; then by completing the square, &c. we get $w =$
$\dfrac{1}{a^2}\sqrt{\dfrac{am^2b^3 + 2apdmb^2 + 3ambpd^2 + 6ap^2d^3 + a^3mbn}{6}}$
$\overline{+ \dfrac{2a^3pdn - nma^2b^2 - 3na^2pd^2 - n^2a^2}{6}} + \dfrac{c^2}{4} - \dfrac{c}{2}$.

EXAMPLE.

Let $a = 1$, $b = 5$, $d = 3$, $m = 5$, $n = 1$, $p = 10$.

First, to find the value of c, the coefficient of w.

$$\frac{n}{6} = .166$$

$$\frac{n}{2a} = .5$$

$$\frac{2pd^2}{a^2} = 180.$$

$$\frac{2mb^2}{3a^2} = 83.33$$

Sum $264 = c.$

Under the vinculum we have,

Positive Terms.			Negative Terms.		
am^2b^3	$=$	3125	nma^2b^2	$=$	125
$2apdmb^2$	$=$	7500	$3a^2npd^2$	$=$	270
$3ambpd^2$	$=$	6750	n^2a^2	$=$	1
$6ap^2d^3$	$=$	16200			
a^3mbn	$=$	.25		Sum	396
$2a^3pdn$	$=$	60			

Sum positive 33660
Sum negative   396

Difference $33264 \div 6 = 5544$, to which add 17424, $(\frac{c^2}{4})$ and extract the square root of the sum, which is found $= 151.552$; this divided by a^2 (1), and $\frac{c}{2}$ taken from it, leaves the value of $w = 19.55$, the weight required.

The above Theorem may be contracted by putting the known quantities, or terms, in the

L

numerator $amb + 2apd - na^2$, $= N = 84$; and those in the denominator, $mb^2 + 3pd^2 + na^2$, $= Q = 396$. Then, by the Theorem in the last Problem, $(\frac{1}{a^2}\sqrt{\frac{QN}{6} + \frac{Q^2}{9}} - \frac{Q}{3a^2})$ we have the solution.

$$\frac{QN}{6} = 5544$$

$$\frac{Q^2}{9} = \underline{17424}$$

their sum $\overline{22968}$, the square root of which $\div$ a^2, or $\times \frac{1}{a^2}$, is 151.55; and $\frac{Q}{3a^2} = 132$, which taken from 151.55, leaves 19.55 as before, but by a much easier process.

On the Maximum of Bodies; when the space is given.

I have now, by various processes, by different methods, and from different principles, demonstrated what the greatest effect is which a given body, acted upon by gravity, can produce in a given *time*. I shall now proceed to demonstrate what the greatest effect is which a given body can produce, in falling through a given *space*. If the ratio of w to x is such, that the effect produced is the greatest at the end of one second of time, it will be the greatest at the end of any other period of time. But if the space is given when the effect is greatest, the ratio of w to x will vary, or be different to what it was when the time was given.

PROBLEM XXXV.

It is required to determine the greatest effect that a given body can produce, in falling through a given space?

Let $w =$ the weight of the body, or power.
$x =$ the resistance, or weight to be raised.
And let them be connected by a line passing over a pulley. (See Fig. II.)

The accelerating force will be $\frac{w-x}{w+x}$ (Prob. I.) which multiplied by 16, gives the feet passed over in the first second. But the time and velocity are as the square root of the space, and will be expressed by $\sqrt{\frac{w-x}{w+x}}$, which multiplied by x, gives $x\sqrt{\frac{w-x}{w+x}}$; which, by the Problem, must be a maximum. Its fluxion is $\overline{2w x \dot{x}-3x^2 \dot{x}}$ $\times \overline{w+x} - \dot{x} \times \overline{wx^2 - x^3} = 0.$ From which we have $2x^3 + 2wx^2 = 2w^2 x$; which, divided by $2x$, gives $x^2 + wx = w^2$; and $x^2 + wx + \frac{w^2}{4} = w^2 + \frac{w^2}{4}.$ By extracting the root on both sides, we get $x = \frac{w}{2}\sqrt{5} - \frac{w}{2}.$

EXAMPLE.

Given $w = 10$; required x?

The square root of $5 = 2.236076$, which multiplied by $\frac{w}{2}$ $(= 5)$, gives 11.18038; from which subtract $\frac{w}{2}$, and there remains 6.18038 $= x$. When $w = 10$, *this number, or value of x, is the maximum for any space, as 4.142 is for any time.*

NOTE 1.—It appears by Prob. III. the time being given, that when the effect is greatest, the power descends through 6.627 feet in one second, and, of consequence, from the laws of falling bodies, through 26.5 in two seconds, and through $3 \times 3 \times 6.627 = 59.6$ in three seconds, &c. Therefore, the weight remaining the same, the space through which it is raised by a given power will be as the square of the time.

NOTE 2.—It appears by this Problem, the space being given, that there is a limited time in which the power will produce the greatest effect.

NOTE 3.—If $T =$ any given time, the number of strokes made in that time will be

$$\frac{\text{T}}{\sqrt{\frac{sw+sx}{rw-rx}}} = \text{T}\sqrt{\frac{rw-rx}{sw+sx}}, \text{ which multiplied by}$$

x, gives $\text{T}\sqrt{\frac{rwx^2-rx^3}{sw+sx}}$ for the maximum; the fluxion of which is the same as that already found in this Problem.

EXAMPLE.

Let $\text{T} = 3600$ seconds, or one hour.

$w = 10.$

$x = 4.142.$

$r = 16.$

$s = 26.5.$

Then $\sqrt{\frac{sw+sx}{rw-rx}} = 2$ seconds, the time of one stroke; and $\text{T}\sqrt{\frac{rw-rx}{sw+sx}} = 1800$, the strokes in one hour; which multiplied by 4.142, the weight raised at one stroke, gives 7456, the weight raised in one hour.

But if $x = 6$;

Then $\sqrt{\frac{sw+sx}{rw-rx}} = 2.57$ seconds, the time of one stroke, by which divide 3600, and we get 1400.7, the number of strokes in one hour; which multiplied by 6, the weight raised by one stroke, gives 8404.2 for the whole weight raised in one hour.

PROBLEM XXXVI.

Given a power w applied to the circumference of a wheel; required the weight x suspended from the axle, when the product arising from the weight into the velocity is the greatest possible at the end of a given space?

Put a = radius of the axle.

     b = radius of the wheel.

     s = the given space through which w descends = 25.

     r = 16.

     x = weight sought.

The accelerating force at w will be $\frac{b^2 rw - baxr}{b^2 w + a^2 x}$.

And as $\sqrt{\frac{b^2 rw - baxr}{b^2 w + a^2 x}} : 1'' :: \sqrt{s} : \sqrt{\frac{wb^2 + sa^2 x}{rwb^2 - rabx}}$ the time in which w falls through the space s; by which divide the time τ, and we have $\tau \sqrt{\frac{wb^2 - abrx}{swb^2 + a^2 sx}}$ = the number of strokes made in the said time. The number of strokes, multiplied by the weight x, which is raised at one stroke, gives $\tau \frac{\sqrt{abwx^2 - a^2 x^3}}{\sqrt{swb^2 + sa^2 x}}$ for the maximum. Its fluxion is $2bwx\dot{x} - 3ax^2\dot{x} \times \overline{b^2 w + a^2 x} - a^2\dot{x} \times \overline{bwx^2 - ax^3} = 0$. From which, by multiplication and reduction, we get $2a^2 x^2 + 3b^2 wx - abwx = \frac{2b^3 w^2}{a}$; and $x^2 + \frac{3b^2 w - abw}{2a^2} \times x =$

$\frac{b^3w^2}{a^3}$. Let $\frac{3b^2w - abw}{2a^2} = n$: then by completing

the square, &c. we get $x = \sqrt{\frac{b^3w^2}{a^2} + \frac{n^2}{4}} - \frac{n}{2}$.

EXAMPLE.

Let $a = 1,\; b = 2,\; w = 10$.

Then $x = 12.74$, and is raised 12.5 feet, while w falls 25 feet.   The time of descent is expressed by $\sqrt{\frac{swb^2 + sa^2x}{rb^2w - baxr}}$, and in this case is 2.38 seconds.

PROBLEM XXXVII.

Let the weight of the wheel be given, $= m$; the centre of gyration y; a power p acting at the circumference: required the weight w suspended from the axle, when the effect of p in raising it through a given space is the greatest possible?

Let the radius of the axle be $= 1$; that of the wheel $= r$.

Then will $\frac{pr^2 - wr}{y^2m + pr^2 + w} \times 16 =$ the number of feet which w ascends in one second.   And as $\sqrt{\frac{pr^2 - wr}{y^2m + pr^2 + w}} : 1'' :: \sqrt{s} : \sqrt{\frac{y^2m + pr^2 + w}{pr - w}} \times \frac{s^{\frac{1}{2}}}{16}$, the time of ascent through the space s; by

which divide the time τ, and the quotient is the
number of strokes made in that time; by which
multiply w, the weight raised at one time, and
we have the whole weight raised in any given
time τ, and which by the Problem must be the
greatest possible: hence we have $\sqrt{\dfrac{prw^2 - w^3}{y^2m + pr^2 + w}}$
for the maximum. ₁The fluxion of which is
$2prw\dot{w} - 3w^2\dot{w} \times \overline{y^2m + pr^2 + w} - \dot{w} \times$
$\overline{prw^2 - w^3} = 0$. From which we get $w^2 +$
$\dfrac{3pr^2 + 3my^2 - pr}{2} \times w = prmy^2 + p^2r^3$; and by
putting $n = $ the coefficients of w, we get $w =$
$\sqrt{p^2r^3 + prmy^2 + \dfrac{n^2}{4}} - \dfrac{n}{2}$.

EXAMPLE I.

Let $p = 10$, $r = 2$, $y^2 = 2$, $m = 1$.

Then $\dfrac{3pr^2 + 3my^2 - pr}{2} = n = 53$; and $w =$
$\sqrt{800 + 40 + 702.25} - 26.5 = 12.77$.

EXAMPLE II.

Let the weight of the wheel be 4, $= m$, and
$y^2 = 6$, to find w?

$n = 86$; and $w = \sqrt{800 + 480 + 1849}$
$- 43 = 12.9$; and the time of rising 25 feet
will be 4 seconds.

EXAMPLE III.

Given the weight of the wheel 10lb.; the power p 10lb.; $y^2 = 10$: required w, and the time of rising 25 feet?

From the above data we find $n = 200$, and $w = \sqrt{800 + 2000 + 10000} - 100 = 13.13$; and the time is $\sqrt{\dfrac{my^2 + pr^2 + w}{pr - w}} \times \sqrt{\dfrac{25}{16}} = 5.9$ seconds.

It appears that, by increasing the weight of the wheel, &c. the time of ascent is prolonged near two seconds, and the weight raised is very little more. (See Prob. XXXII.)

PROBLEM XXXVIII.

Let the power w be applied to the axle, to raise the weight x suspended from the wheel (the rest of the notation as in the last Problem) through a given space; required x, when the effect is the greatest?

$\dfrac{w - rx}{y^2m + r^2x + w}$ is the space passed over by w compared with gravity (1); and, by the last Problem, $\dfrac{wx^2 - rx^3}{y^2m + w + r^2x}$ is the greatest effect. Its fluxion is $2wxx - 2rx^2x \times \overline{y^2m + w + r^2x}$

M

$$- r^2x \times \overline{wx^2 - rx^3} = 0. \quad \text{From which we find}$$

$$x = \sqrt{\frac{2w^2 + 2mwy^2}{2r^3} + \frac{n^2}{4}} - \frac{n}{2}.$$

Note. $\dfrac{3rw + 3my^2 - r^2w}{2r^3} = n.$

NOTE.—In constructing machines, it will, in general, be of more importance to compute the effect produced by a body in passing over a given space, than at the end of a given time.

But suppose it should be required to assign the distance at which a given power must be applied from the axis of a wheel, so as to produce a greater number of revolutions, in falling through a given space, than could be produced by any different application of the said power, in falling through the same space, in any proportionable part of the time: the answer is, apply the power as near the axis as strength in the machine and other circumstances will admit. (See Prob. XIV and XV.) It has been proved, that if a body falls through a given space, suppose 20 feet, in 20 seconds, and produces 100 revolutions, if applied at twice the distance from the axis it will fall through the same space in 10 seconds, but will only produce 50 revolutions, the *vis inertiæ* of the power not considered; but if it is taken into the computation, it will require more than 10 seconds to make 50 revolutions.

For example,

Let m = the weight of the wheel = 31.9.
    r = the centre of gyration = 10.
    p = the power = 10.
    x = the distance from the axis at which
        p is applied = 1.

Then $\frac{16px^2}{mr^2 + px^2}$ = the feet fallen through by p in the first second = $\frac{1}{20}$. And as $\frac{1}{20}$ foot : $1^{2\prime\prime}$ :: 20 feet : $20^{2\prime\prime}$, the time of descent. But if $x = 2$, by the same process, the time of descent will be found = 10.046 seconds; and in falling through a given space, the number of revolutions must be inversely as the diameters. Hence, by increasing the distance at which the power is applied, we diminish the effect when the space is given. But if the time was given, the distance might be found such, that the effect would be the greatest.

PROBLEM XXXIX.

Given the weight of a fly = m = 31.9; the centre of gyration = r = 10; the power = p = 10: required x, the distance at which p

must be applied, to produce, in a given time, the greatest number of turns?

$\frac{mr^2}{x^2}$ = the whole resistance of the wheel reduced to p, to which if we add p, we have $\frac{mr^2}{x} + p$ = the whole mass; the moving power being divided by this, gives $\frac{px^2}{mr^2 + px^2}$, which multiplied by 16, gives the space passed over by p in the first second; which divided by $2x \times q$ $(2x \times 3.1416)$ gives $\frac{px}{2q \times \overline{mr^2 + px^2}}$, the number of revolutions performed in that time. The fluxion of which is $px \times \overline{mr^2 + px^2} - 2pxx \times px = 0$. From which we get $pmr^2 + p^2x^2 = 2p^2x^2$; and $x = r\sqrt{\frac{m}{p}} = 10\sqrt{\frac{31.9}{10}} = 17.8$, which, in this case, is the distance at which the power ought to be applied, in order to produce the greatest number of revolutions in a given time.

Given the weight of a beam suspended by the middle, $= m = 200$lb.; and a weight $p = 100$, suspended from one end: required the weight to be suspended from the other, so that the effect may be the greatest after having ascended through a given space?

It is presumed that the ends of the beam are circular, and that the weights move in perpendicular directions.

The accelerating force will be $\dfrac{p-x}{\frac{m}{3}+p+x}$ (see Prob. XXVII & XXXII); the square root of which multiplied by x, gives the maximum $=$ $\sqrt{\dfrac{px^2-x^3}{\frac{m}{3}+p+x}}$. Its fluxion is $2pxx - 3x^2x \times \overline{\frac{m}{3}+p+x}$ $- x \times \overline{px^2-x^3} = 0$. From which we have $x^2 + px + \dfrac{mx}{2} = \dfrac{pm}{3} + p^2$. Let $p + \dfrac{m}{2} = n$; then, by completing the square, &c. we find x $= \sqrt{\dfrac{pm}{3}+p^2+\dfrac{n^2}{4}} - \dfrac{n}{2}$.

On Steam Engines.

PROBLEM XLI.

It has been demonstrated, that when the pressure upon the steam-piston is to the weight of the pump-rods and water as 1000 to 618, the effect is the greatest when the piston descends. And if it is so adjusted, that the weight of the rods, &c. when in the water, be to the weight of the steam-piston, plug frame, &c. as 1000 to 618, the engine will do the most work

in a given time.   If the pressure of the air is
taken at 14lb. upon an inch, then as 1000 : 618
: : 14 : 8.65, the weight with which the end
over the shaft should be loaded for every square
inch of the steam-piston ; (not considering the
friction of the parts, or weight of the beam,)
which would amount to 1245lb. for every square
foot of the steam-piston : but every thing considered, this is far too much, and we may, in
general, make the load from 800 to 860lb.
per foot.   In large engines, the friction is less,
in proportion to their power, than in small
engines; of consequence the load may be more.

Put $d =$ diameter of the steam-cylinder, in feet.
   $q = .7854.$
   $l =$ the length of the stroke, in feet.
   $p =$ the pressure per foot, in pounds avoir-
         dupoise.
   $w = 62.5$lb. weight of a cubic foot of water.
   $h =$ the number of feet the water is to be
         raised.
   $c =$ diameter of the pump, in feet.
   $g = 6.1276$, ale gallons in a cubic foot.
   $n = 10.28$, cubic feet in a hogshead.

   Then will

   $qd^2 =$ area of the steam-piston, in feet.

pqd^2 = the weight of the water to be lifted, in pounds avoirdupoise.

$\frac{pqd^2}{w}$ = cubic feet of water in the pump; which divided by the height h, gives

$\frac{pqd^2}{hw}$ = the area of the pump-piston; which multiplied by the length of the stroke l, gives

$\frac{lpqd^2}{hw}$ = cubic feet raised per stroke.

From which equations we have the following Theorems:—

Theo. 1. $d = \sqrt{\frac{whc^2}{p}}$.

Theo. 2. $c = \sqrt{\frac{pd^2}{hw}}$.

Theo. 3. $h = \frac{pd^2}{wc^2}$.

Theo. 4. lqc^2 = cubic feet per stroke.

Theo. 5. $glqc^2$ = gallons per stroke.

EXAMPLE I.

Given the diameter of the steam-cylinder, 3 feet; the length of the stroke, 6 feet; the depth of the well, 50 feet: required the diameter of the pump, and the quantity of water raised at one stroke?

Here we have given the diameter of the cylinder $= 3 = d$; the length of the stroke $= 6 = l$; the depth of the well $= 50 = h$; to find c; which, by Theo. 2, is $= \sqrt{\frac{pd^2}{hw}} = \sqrt{\frac{7515}{3125}} = 1.55$ feet. And the quantity per stroke is, by Theo. 4, $= lqc^2$, $= 6 \times .7854 \times 2.404 = 11.328$ cubic feet.

In the above, p is taken at 835lb. instead of 1245, which it ought to be taken at, were it not for the various sorts of resistance, friction, &c. which do not enter into the above computation.

If all cylinders were equally well bored, and the work executed in the same manner, an estimate might be made for friction, &c. which, compared with the moving power, would always be, in the cylinder, as the circumference to the area of the piston: or, in different engines, it would be as the diameters of the cylinders and pumps. To which should be added the weight of the beam, rods, plug-frame, &c.

EXAMPLE II.

Given the diameter of the pump, 1 foot; the height to which the water is to be raised, 240 feet; to find the diameter of the steam-cylinder?

Here we have $w = 62.5$lb.; $h = 240$; $c = 1$: and $whqc^2 = pqd^2 = $ the whole weight of the cylinder of water to be lifted. From which we find, Theo. 1, $d = \sqrt{\dfrac{whc^2}{p}} = \sqrt{\dfrac{15000}{835}} = \sqrt{17.96} = 4.23$ feet, the diameter required.

EXAMPLE III.

Given the quantity of water to be raised in one hour, 200 hogsheads; the depth of the water, 100 yards; length of the stroke, 6 feet; number of strokes in a minute, 10: required the diameter of the cylinder and pump?

200 hogsheads, the quantity raised in one hour, is equal to 12600 gallons, which divided by 600, the number of strokes made in the same time, quotes 21, the gallons to be raised at one stroke. Therefore, (Theo. 5) $21 = glqc^2$, and $c = \sqrt{\dfrac{21}{glq}} = .852$ feet, or 10.22 inches, the diameter of the pump: and (per Theo. 1) $d = \sqrt{\dfrac{whc^2}{p}} = 4.04$ feet, the diameter of the cylinder.

N. B. If the weight of the pump-rods and plug-frame be taken into the computation; let their weight, when the weight of the steam-piston is taken from them, be put $= n$ pounds.

N

Then will $d = \sqrt{\dfrac{whc^2}{p} + \dfrac{n}{qp}}$ And $c =$ $\sqrt{\dfrac{pd^2}{hw} - \dfrac{n}{qhw}}$.

EXAMPLE IV.

Required the dimensions of an engine to raise 250 hogsheads per hour to the altitude of 55 yards, the weight of the rods being 2000lb.?

$250 \times 63 = 15750$, the gallons per hour, which, at 12 strokes in a minute, is 22 gallons per stroke $= 3.59$ cubic feet $= qlc^2$; and $c^2 = \dfrac{3.59}{ql}$; the square root of which is .87 feet, or 10.4 inches, the diameter of the pump. And $d = \sqrt{\dfrac{whc^2}{p} + \dfrac{n}{pq}} = 3.53$ feet, the diameter of the cylinder.

PROBLEM XLII.

Given the length of an engine beam $= 2a$, its weight $= m$, weight of each horse head, or arch at the end, $= c$, weight of the steam-piston and chain $= s$, weight of the plug-frame and its arch $= u$, its distance from the centre $= t$, pressure of the atmosphere upon the piston $= p$; required the weight of water lifted, or resistance overcome, when the effect is greatest?

Suppose the beam to be suspended by the centre, and that the piston and chain are balanced by a part of the pump-rod and chain at the other end: then will the whole weight of the beam, arches, piston, &c. be $2c + 2s + m$, which put $= r$, and let the centre of gyration in the beam be y, $= \sqrt{\dfrac{6cb^2 + 6sa^2 + ma^2}{6 \times c + s + ,3m}} = \sqrt{\dfrac{a^2}{3} \times \dfrac{4c + 4s}{2c + 2s + m} + 1}$, then will $\dfrac{ry^2}{a^2}$ be the whole resistance of the beam, &c. reduced to the end to which the power is applied.

Let $x =$ the resistance sought: then will $\dfrac{p - x \times 16}{p + \dfrac{ry^2}{a^2} + x}$ be the space passed over in the first second.

Let $l =$ the length of the stroke: then,

As $\dfrac{p - x \times 16}{p + \dfrac{ry^2}{a^2} + x} : 1'' :: l : l \times \dfrac{p + \dfrac{ry^2}{a^2} + x}{p - x \times 16}$; the square root of which is the time of passing over the space l. And the greatest effect will be expressed by $\sqrt{\dfrac{px^2 - x^3}{p + \dfrac{ry^2}{a} + x}}$; the fluxion of which is $\overline{2px\dot{x} - 3x^2\dot{x}} \times \overline{p + \dfrac{ry^2}{a^2} + x} - \dot{x} \times \overline{px^2 - x^3}$ $= 0$. From which we get $x^2 + \overline{p + \dfrac{3ry^2}{2a^2}} \times x = p^2 + \dfrac{pry^2}{a^2}$. Let $p + \dfrac{3ry^2}{2a^2} = n$, the coefficients of

x; and we shall have $x^2 + nx + \dfrac{n}{4} = \dfrac{pry^2}{a^2} + p^2$ $+ \dfrac{n^2}{4}$: from which we find $x = \sqrt{\dfrac{pry^2}{a^2} + p^2 + \dfrac{n^2}{4}}$ $- \dfrac{n}{2}$. Then, to find the additional weight of the pump-rod, so that the ascent of the piston may be performed in the same time as the descent, we have As $x : p :: s : \dfrac{sp}{x} =$ the whole weight of the pump-rods and chain, from which subtract s, and we have the additional weight sought; which, if the rods be not heavy enough, must be supplied by an equivalent weight added to the end of the beam, and must be further increased by reducing the plug-frame to the end of the beam, in the following manner, viz. $\dfrac{t^2u}{a^2}$ $=$ the resistance at the end of the beam; and As $x : p :: \dfrac{ut^2}{a^2} : \dfrac{put^2}{xa^2}$; hence $\dfrac{sp}{x} + \dfrac{put^2}{xa^2} =$ the whole weight suspended at the pit end of the beam. This diminished by s, and the remainder taken from x, leaves $x - \dfrac{sp - sx}{x} + \dfrac{put^2}{xa^2} =$ the weight of water to be lifted, not considering the friction.

EXAMPLE I.

Let $a = 12$, $m = 1000$, $c = 280$, $s = 700$, $u = 200$, $t = 7$, $p = 2000$: then will $r =$

93

2960, and $y^2 = 111.55$; also $\frac{ry^2}{a^2} = 2293$; $n = 5440$.

$$\frac{pry^2}{a^2} = 4586000$$

$$p^2 = 4000000$$

$$\frac{n^2}{4} = 7398400$$

Sum 15984400; the square root of this is 3998, from which subtract $\frac{n}{2}$ ($= 2720$) and we have $1278 = x$, the whole resistance, which must be diminished by $\frac{sp}{x} - s + \frac{put^2}{xa^2} = 502$: and we have 785 for the weight of the water to be lifted. And if $l = 6$ feet, the time of descending will be $\left(\sqrt{l \times \dfrac{p + \frac{ry^2}{a^2} + x}{p - x \times 16}}\right)$ 1.6 seconds, which doubled, gives the time in which the piston, &c. ascends and descends $= 3.2$ seconds, which would give 18 strokes in one minute, if the piston made no stop at the top or bottom.

EXAMPLE II.

Let $a = 12$, $m = 5000$, $c = 1000$, $s = 2000$, $u = 500$, $t = 7$, $p = 14250$: then $2c+2s+m = 11000 = r$; also $\frac{a^2}{3} \times \frac{4c + 4s}{2c + 2s + m} + 1 = 100.3636 = y^2$; and $\frac{ry^2}{a^2} = 7666\frac{2}{3}$. $\frac{3ry^2}{2a^2}+p = 25750 = n$: from which we have

$$\frac{pry^2}{a^2} = 109250000$$

$$p^2 = 203062500$$

$$\frac{n^2}{4} = 165765625$$

Sum 478078125; its square root is 21865, from which take $\frac{n}{2}$, and we have $9990 = x$; from which take $\frac{sp}{x} - x + \frac{put^2}{xa^2} (= 1095)$ and we have the weight of water $= 8895$lb.

In the first Example, every square foot of the piston raises 1288lb.; and in the second, 1413lb. In the first, the strokes per minute are 18.75; in the second, 17.96; which, multiplied by the weight raised at one stroke, gives 25377 and 24150, which difference arises from the superior weight of the beam, &c. in the second Example.

Had the value of x been sought, as in Prob. XXXV. neglecting the weight of the beam, &c. in the first Example it would be found 1236, and the strokes per minute would be 24, hence the weight raised per minute $= 29664$lb.; in the second Example, the weight 'per stroke, and the strokes per minute, would be the same as in the first. Hence, computations on the power of engines, or machines of any kind, must be erroneous when the weight or resistance of the machine is neglected.

When an engine is employed to raise a large quantity of water to a little height, the spear, or pump-rod, is not heavy enough to lift the piston with a due velocity; hence it has been common to load that end of the beam: but it would certainly be an advantage to make the end over the pump something longer, so that by its increased weight it may bring up the piston with a proper velocity, and also make a longer stroke in the pump, and of consequence bring up a greater quantity of water (see Prob. XIX, XX, and XXI). The theory considers every part of an engine perfect; but as this can scarcely ever be true, it may be an advantage to make the load less than the theory assigns, in order that it may strike faster, by which means the loss of water by the piston, through the valves, &c. will in part be diminished. And if the method of computing the power in the last Problem should be thought too intricate, an engine constructed by the Theorems in Prob. XLI. will, if well executed, be pretty near the maximum.

A

Treatise on Mills.

PART III.

ON THE VELOCITY OF FLOWING WATER.

To determine the quantity of water discharged in a given time through given apertures, and at different depths below the surface,
is, in many cases, of great importance. Sir
Isaac Newton, in his *Principia*, Book 2, Theo.
8, Prob. 36, has demonstrated that the velocity
of water flowing through holes in the bottom
or side of a vessel, ought to be equal to the
velocity which a heavy body would acquire in
falling through a space equal to the distance
between the surface of the water and the place
where it is discharged.   Hence, at the depth
of 16 feet, a stream of 32 feet in length ought

to flow out in one second of time. And from
the laws of falling bodies, it follows, that as
the square root of 16, is to the velocity of the
stream flowing out at that depth; so is the
square root of any other depth, to the velocity
at that depth: that is, the expence of water
through equal apertures is directly as the square
roots of their depths. But Sir Isaac, in making
experiments, found the velocity thus determined to be too great, which in different cases he
corrected. The friction against the sides of the
hole, and the oblique direction of the particles
of water before they reach the hole, both tend
to diminish the velocity of the stream. And
if these causes could be removed, especially the
latter, the *Newtonian theory* would certainly
be confirmed by experiment; or rather, experiment would exactly agree with theory. If
we suppose water running into the top of a tube
of equal diameter, and that there is no attrac-
tion, or friction, between the particles of water
and the said tube, the velocity of the water, or
of each particle at the bottom, will be the same,
or equal to that which they would have acquired in falling through the same space, without the tube, toward the earth. Hence to
obtain the true velocity, under different circumstances, we must correct the computed velocity
by experiments. And in order to satisfy my-

self of the accuracy of those made by others,
I have taken the trouble of making a consider-
able number; a few of which I will insert.

From the experiments made by some authors,
it appears, that the velocity of effluent water is
much less than assigned by the Newtonian theory: so much less, that they have concluded the
velocity ought to be the same as that acquired
by a body falling through a space equal to half
the depth. But this must be erroneous; as the
observed, or real velocity, can never exceed that
assigned by a just theory, whereas it does exceed
that assigned by this theory, but from impediments, may be expected to fall short, more or
less. When I began to make experiments, I
wished, as much as possible, to get clear of these
obstructions, &c.; and, from the recommenda-
tion of authors, I began with making the apertures in plates about $\frac{1}{20}$th of an inch thick: the
result of which is contained in the following
Table, where it appears that the observed velocity, at the depth of 8 feet, falls $4\frac{1}{4}$ feet short
of the computed.—I afterwards tried various
forms, all of which gave more water than the
thin plate. But as my view was to obtain the
greatest velocity, I shall take no notice of any
but the cone, which I found to be the best.
This I made of brass, about $\frac{1}{10}$th of a foot in

length, and the top diameter about twice as
much as the bottom, or aperture. The stream
discharged through this is remarkably different
from that discharged through the plate. This
appears like a piece of crystal glass, to the distance of some feet, more or less, according to
the head; neither has it that contraction just
below the aperture as observed in one flowing
through a thin plate; and the observed velo-
city at the depth of 7 feet, only falls $1\frac{1}{4}$ foot
short of the computed.

PROCESS.

The aperture was first bored near the intended size; it was then finished with a piece of hard
steel wire, one end of which was made like a
square pyramid, and the other fixed in the spindle of a lathe. The wire was afterwards broken
into a number of pieces, which were placed
close to each other on a diagonal scale, and the
sum of their diameters observed; they were
also measured by a screw micrometer; all the
different pieces were measured by several gentlemen, who never differed so much as $\frac{1}{1000}$th part
of an inch in the diameter of a piece. And from
the velocity of the water through the different
holes, or those made by different wires, I have
reason to conclude that their diameters were all
measured to the greatest degree of accuracy.

To obtain the quantity discharged in a given time.—An assistant closed the aperture with a finger till the tube was full, and held the pendulum of a half-second time-piece with the other hand; then, removing both hands at the same instant, the water was suffered to flow as long as convenient, (by a constant supply, the head of water was always the same during the experiment) and was stopped at the end of any given time, to the greatest degree of accuracy.

To find the velocity.—The water discharged was weighed, reduced into cubic inches, and divided by the area of the aperture, in inches; by which the length of the stream run out in the whole time was known.

Experiments, with the aperture made in a plate about $\frac{1}{40}$th of an inch thick.

TABLE. I.

Number.	Head, in feet	Area of the hole, in inches.	Weight run out in a minute, in oz avoirdupoise	Velocity per second, by computation.	Velocity per second, by experiment	Difference at the depth of 16 feet.
1	1.375	.0055	16.97	9.38	7.4	
2	2.2	.0055	21.4	11.86	9.3	
3	4.5	.0055	30.6	16.9	13.35	6.7
4	8.	.0055	40.8	22.62	17.8	
5	1.375	.01227	37.94	9.38	7.42	
6	2.2	.01227	48.1	11.86	9.41	
7	4.5	.01227	68.4	16.9	13.37	6.8
8	8.	.01227	92.2	22.62	18.1	
9	3.66	.0046	23.17	15.31	12.08	
10	8.	.0046	34.39	22.62	18.	6.6
11	3.66	.007	35.36	15.31	12.12	
12	8.	.007	90.9	22.62	18.08	6.5
13	3.66	.09	454.	15.31	12.106	
14	8.	.09	674.	22.62	17.97	6.6

In the six following Experiments, the Water passed through a cone $\frac{1}{10}$th of a foot in length.

TABLE I. CONTINUED.

Number.	Head, in feet.	Area of the hole, in inches.	Weight run out in a minute, in ounces avoirdupoise.	Velocity $\mathscr{P}$ second, by computation.	Velocity $\mathscr{P}$ second, by experiment.	Difference at the depth of 16 feet.
15	3.61	.0046	26.87	15.2	13.9	
16	3.61	.007	41.16	....	14.23	2
17	3.61	.0141	85.3	....	14.52	
18	7.	.0046	37.43	21.16	19.53	
19	7.	.007	57.4	....	19.68	2
20	7.	.0141	117.	....	19.91	

The following Experiments were made by the Abbé Bossuet: his reservoir was 3 feet square within; the orifices through which the water issued were made in brass plates, about $\frac{1}{24}$th of an inch thick.

N. B. The measures are taken by the Paris foot.

Height of the Water above the centre of the hole, 11 feet 8 inches and 10 lines.

TABLE II.

By a circular orifice, 6 lines in diameter, 2311 ⎫
   1 inch   do.   9281 ⎬ cubic inches $\mathscr{P}$ minute.
   2 inches   do.   37203 ⎭

By a square orifice, 1 inch by 3 lines .. 2933 ⎫
                1 inch by 1 inch .. 11817 ⎬ cubic inches
                2 inches square   .. 47361 ⎭ ∜ minute.

Height of the Water, 9 feet.

By a circular orifice, 6 lines in diameter, 2018 ⎫ cubic inches
             1 inch   do.   8135 ⎭ ∜ minute.

Height of the Water, 4 feet.

By a circular orifice, 6 lines in diameter, 1353 ⎫ cubic inches
             1 inch   do.   5436 ⎭ ∜ minute.

In English measure, the head to the six first experiments is 12.5104 feet ; the computed velocity is 28.296 feet; and the observed, at a mean, is near 17½ feet.

———◆———

Experiments from F. D. Michelotti's *Sperimenti Idraulici*, 1767. The reservoir was 20 feet high, and 3 feet square within, and had openings at different distances below the top.   It was supplied by a canal 2 feet wide, the bottom of which was horizontal.   The water ran into a cistern, whose area was 289 square feet, of an uniform figure, and the quantity was measured by taking its height in this cistern.—The measures are French.

TABLE III.

Size and kind of hole.	Head above the centre of the hole.				Time of running.	Cubic feet of water expended.		
	Feet.	Inch.	Lin.	Pa.	Minutes	Feet.	Inch.	Lin.
	6	7	4	3	10	463	7	3
	6	10	2	8	12	566	5	6
Square, of	11	8	1	6	8½	516	9	5
3 inches.	11	9	9	10	10	612	1	5
	21	8	3	6	5	415	5	3
	21	8	7	0	6	499	2	8
Square, of	6	7	6	0	15	329	9	8
2 inches.	11	5	1	4	15	423	5	7
	21	5	3	7	10	385	4	0
Square, of	6	9	1	0	30	158	6	7
1 inch.	11	10	8	1	24	163	9	6
	21	6	1	0	60	562	11	4
Circular of	6	8	4	0	15	542	10	6
3 inches	11	7	1	0	12	570	11	8
diameter.	21	7	4	0	8	521	3	7
Circular of	6	9	5	0	30	488	8	3
2 inches	11	8	8	0	28	589	6	5
diameter.	21	10	10	0	20	575	5	10
Circular of	6	10	6	0	60	247	4	3
1 inch	11	8	11	0	60	324	1	5
diameter.	22	0	2	0	60	444	6	5

In the first Experiment, the head, reduced to English measure, is 7.049 feet, and the computed velocity at 21.2 feet. The quantity run out in ten minutes, in French measure, is 463.604 cubic feet; the area of the hole is 0625 feet, by which divide the quantity, and we have 7416.06, the feet run out in ten minutes, which is equal to 12.36 French, or 13.175 English feet, in one second: or, the true quantity is to the computed, nearly as 5 to 8 : and nearly the same result is drawn from all the Experiments made by this author, of which I have copied but a few.

Experiments by Messrs. Brindley & Smeaton : the water extended over a large surface, the depths measured from that surface to the top of the holes.

TABLE IV.

Size of the hole.	Head, in feet.	Time in which 20 cubic feet run out.
	1	9 min. 22 sec.
	2	6 .. 40 ..
1 inch square.	3	5 .. 20 ..
	4	4 .. 44 ..
	5	4 .. 14 ..
½ inch square.	6	17 .. 33 ..

TABLE V.

Notches 6 inches wide.	Time in which 20 cubic feet run out.
1 inch deep	7 min. 16 sec.
1⅜	4 .. 55 ..
2 5/16	2 .. 19 ..
3⅛	1 .. 33 ..
6¼	0 .. 30 ..
5	0 .. 46 ..
1¼	5 .. 26 ..
1⅝	3 .. 55 ..
5⅝	0 .. 42 ..

Experiments by Professor Helsham, of the
University of Dublin.

TABLE VI.

Head, in feet.	Diameter of the hole, in inches.	Time, in seconds.	Weight, in grains.
4	.1	10	2944
4	.4	..	47040
4	.5	..	72960
4	.8	..	178560
2	.1	..	2087
2	.4	..	33600
2	.5	..	51840
2	.8	..	128400

In the first fourteen Experiments in Table I.
the velocity, or quantity discharged, is something greater than in the following Tables,
though the orifices were made in flat plates:
this is owing to the form of the lower part of
the tube being the frustum of a cone, and the
plate being half an inch in diameter. In the
six last experiments, where a cone is substituted
for the plate, we have the velocity considerably
increased: so much as, at the depth of 16 feet,
to fall only 2 feet short of the computed velocity. But if the laws of hydrostatics assigned
no greater velocity to flowing water, than that
which would be acquired by a heavy body in
falling through half the depth, we should have
an effect produced without a cause. For the
velocity at the depth of 16 feet should be, ac-

P

cording to that law, the same as a body would acquire in falling through 8 feet, or 22.627 feet per second; whereas it is, in fact, 30 feet. And whatever advantages I may have taken, in respect to the form of the containing vessel, this alters not the pressure of the fluid, or the laws of nature, but tends to remove the impediments.

But from these Experiments, we are not to conclude, that the velocity of water, as flowing through orifices made in large surfaces, is in the same ratio. Sir Isaac Newton concluded, that the observed, or real velocity, was less than the computed, in the ratio of 1 to the square root of 2, or as 1 to 1.414; the Abbé Bossuet, as 100 to 150; and F. D. Michelotti, as 5 to 8, &c.— The quantities in the foregoing Tables, reduced to the same denomination, stand as below:—

$$
\begin{array}{rl}
\text{Newton} & .707 \\
\text{Bossuet} & .615 \\
\text{Banks} & .750 \\
\text{Michelotti} & .625 \\
\text{Helsham} & .705 \\
\text{Smeaton} & .631 \\
\hline
6) & 4.033 \\
\hline
\text{Mean} = & .672
\end{array}
$$

Bossuet and Michelotti have made Experiments upon the largest scale of any I have heard of, though they do not give the greatest velocity; yet they were both agreed with Sir Isaac in the method of computing it: hence, they could not be biassed to favour a different theory. If we take the mean velocity of the above Experiments, we find it to be $\frac{672}{1000}$ of the computed; or their ratio is as $5\frac{3}{8}$ to 8 : viz. at the depth of a foot, the actual velocity from them is $5\frac{3}{8}$ feet, and in the same proportion for any other depth.

EXAMPLE.

Given the depth 6 feet, to find the velocity?

Let d = the depth in feet.

$\qquad v$ = the velocity required.

Then, $.672 \times 8 \sqrt{d} = v$; $\sqrt{d} = 2.449$, which $\times 5\frac{3}{8} = 13.163$, the required velocity.

When water is discharged through perpendicular sections, the velocity at the bottom will be something greater than at the top of the orifice; but if the depth or breadth of the orifice be but little, compared with its depth below the surface of the dam, we may take the depth of the centre of the hole for the mean depth, without sensible error.

RULE.

Measure the depth in feet, extract the square root of that depth, and multiply it by 5.4, which gives the velocity in feet per second : this multiplied by the area of the orifice in feet, gives the number of cubic feet which flows out in one second.

EXAMPLE.

Let the depth be 10 feet, the breadth of the orifice 7 inches, and its horizontal length 4 feet; required the quantity of water expended in one second?

The square root of 10 is 3.162, which multiplied by 5.4, gives 17.0748 feet, the velocity per second.   The area of the orifice is $\frac{4 \times 7}{12} =$ $2\frac{1}{3}$ feet, by which multiply the velocity, and we have 39.84 cubic feet per second.

If the breadth of the orifice is great, compared with the head, we must compute the velocity at the top and bottom, and find the area of a perpendicular section discharged in a given time.

EXAMPLE.

Given the perpendicular depth of the orifice 2 feet, its horizontal length 4 feet, and its top 1 foot below the surface of the water in the dam, to find the quantity discharged in one second?

See Fig. XX, where B is the surface of the water, BA the depth of the upper side of the orifice, and BD the depth of the bottom thereof. BD is the axis of a parabola, AC and DG ordinates which represent the velocity of flowing water at these distances from the vertex; we have, therefore, to find the area of the section ACGD, which will be done by finding the area of BDG, and deducting that of BAC from it.

Put DB $= d$, AC $= a$, DG $= n$, BA $= c$. Then the area of BGD $= \frac{2dn}{3}$, and that of BAC $= \frac{2ca}{3}$; hence $\frac{2dn}{3} - \frac{2ca}{3} =$ area of ACGD. For, by the above rule, the velocity at the top is 5.4, at the bottom 9.24, and the area required $= 14.82$, which multiplied by the length, gives 59.3 cubic feet per second.

Of the Quantity of Water discharged through Slits or Notches cut in the side of a vessel or dam, and open at the top.—See **Fig. XXI.**

If the surface is considered without motion, and the velocity below increasing with the square root of the depth, then the area of the perpendicular section will be a part of a parabola, and will be found by multiplying the velocity at the bottom by the depth, and taking two-thirds of the product for the area, which

again multiplied by the breadth of the notch,
gives the quantity or number of cubic feet discharged in the given time.

For example,

Let the depth of the notch be 5 inches, and
the breadth 6 inches; required the quantity
run out in 46 seconds?

The depth in feet is .4166, its square root is
.6455, which multiplied by 5.4 × $\frac{2}{3}$ = 3.6,
gives 2.3238 ; this multiplied by .4166, the
depth, produces .96825, which being multiplied
by the breadth, half a foot, or .5, gives .48412
feet per second; this multiplied by the time, 46
seconds, gives 22.269 cubic feet.—The sixth
Experiment in Table V. gives 20 cubic feet.

For another example,

Let the depth be 3$\frac{1}{8}$ inches, the breadth 6
inches; required the quantity which flows over
in 93 seconds?

Let d = the depth in feet.
    b = the breadth in feet.
    n = 3.6.

Then, 3.6 $db\sqrt{d}$ = cubic feet per second: in
this example, d = .2604, b = .5, $\sqrt{d}$ = .51;
these multiplied together, produce 22.227 cubic

feet; more than the experiment by $2\frac{1}{4}$ feet, nearly.—At the depth of 1 inch, the computed quantity is 18.727 cubic feet, but by the experiment it is 20. Also, through a notch 1 inch square, the quantity, as near as can be measured by experiment, is 8.6 cubic feet, and by computation only 7.3. From which comparisons it appears, that all the experiments have not been equally accurate. In the above method of computing, the surface of the water is supposed to be without motion; this is not strictly true in any case, but more especially in rivers; though the deviation from experiment, in still water, is not great. And if we wish to take the velocity of the surface into the computation, we may proceed as in finding the quantity which flows through an orifice just below the surface.

Let $v =$ the velocity of the surface: then will $\overline{\frac{v}{5.4}}\Big|^2 = h$, the head which would produce that velocity; if this is added to the depth of the notch, we shall have the following Theorem, viz. $\frac{2}{3} \times \overline{d + h} \times 5.4 \sqrt{d} - \frac{2hv}{3} =$ the area of the section, which, multiplied by the breadth of the section, gives the number of cubic feet per second.

EXAMPLE.

Let $d = 1$ foot, the depth of the notch; $b = 1$ foot, the breadth; $v = 1$ foot per second; $h = \overline{\frac{v}{5.4}}\Big|^2 = .034.$

Then $\overline{d + h} = 1.034$, $\sqrt{d} = 1$, and $\frac{2hv}{3} = .0228$; or $1.034 \times 5.4 \times \frac{2}{3} - 1 \times \frac{2}{3} = 3.7003$, the area; or $\overline{d + h} \times 3.6\sqrt{d} - \frac{2hv}{3} =$ the area, which multiplied by b, the breadth, gives 3.7 cubic feet per second.

In this example, by taking the velocity of the surface to be one foot per second, the quantity run out in a second is one-tenth of a cubic foot more than it would have been had the surface been supposed without motion.

On the impulsive Force of effluent Water.

The pressure of water upon a given surface is directly as the depth, whatever may be the figure or magnitude of the containing vessel or reservoir: but the velocity is as the square root of the depth, and the impulse as the square of the velocity, or as the depth. In addition to the demonstration already given, I shall add the following observations:

1. The depth of the water may be considered as the power which projects or impels the flowing stream.

2. The quantity and velocity conjointly are the effect of that power.

3. Could we consider the velocity alone, and suppose the power to act upon a single particle, in that case, twice the depth would produce a double velocity in the said particle.

4. Also, if a given power communicates a certain velocity to one particle, double that power would communicate the same velocity to two particles.

5. Therefore, if a given power gives a certain velocity to one particle, it will require four times that power to communicate two degrees of velocity to two particles. And in water flowing through a given hole, the quantity will increase directly with the velocity, viz a double velocity must give a double quantity, and of consequence the power which produces it is always as the square thereof. From hence we infer, that the effect of a stream of water flowing through a given orifice, ought to be directly as the depth of that orifice: that is, if a stream of water flows through a hole at the depth of one

foot, and produces a given effect, a stream flowing through the same hole, at the depth of four feet, ought to produce four times the effect, though the expence of water be but double.

The absolute force of a stream has commonly been estimated by the weight of a column or prism of water, the section of which is equal to the area of the orifice: this is the true measure of the pressure against a given surface, but the water in motion ought to produce a greater effect.

In order to compare the impulse of a given stream with the weight of the column which impels it, I made the experiments in the following Table.

FIG. XXII.

The lever or steel-yard N L, moveable on its centre c, was so fixed, that a point in the plate N (the head of a hook *h* riveted into it) was exactly under the stream *s*, and the moveable weight w shifted till the force of the stream would raise the end L from the prop P. When the stream was stopped, or the lever removed, weight was applied to the hook *h*, till the end L was raised from the prop, as by the stream; by which means the force of the stream, compared with a dead weight, was obtained.—In the third

column of the Table, we have the weight of a
cylinder of water, the length of which is equal
to the depth in the first column, and the section
equal the area of the hole in the second; and
while the hole remains the same, the weight
must be exactly as the depth; and also in the
fourth column, the force, though more than
the weight, is still directly as the depth; or,
the forces, as measured by experiments, though
they differ a little, are near enough to prove
that they increase with the square of the velocity, or with the depth, directly.

TABLE.

Depth, in feet.	Area of the hole.	Weight of the perpendicular column.		Force on the steel-yard.	
		Dwts.	Grs.	Dwts.	Grs.
1.5	.0055	0	25	2	0
2.0833	.0055	1	11	2	18
8.	.0055	5	13½	10	10
1.25	.02259	3	14	6	0
1.5416	.02259	4	10	7	6
8.	.02259	22	21	38	6
2.375	.0141	4	0	7	0
3.458	.0141	6	4	10	6
7.0416	.0141	12	13½	21	10
9.3125	.0141	16	14.8	28	5

But the force of a stream, or rather the impulse, cannot be measured by its effect on a
wheel, when that wheel is confined in a channel;
for in that case, the stream, when it has struck
the float-boards, cannot make its escape, and
is therefore heaped or raised against the floats
much higher than it would be if at liberty to
pass.

On Emptying of Vessels.

The velocity with which water flows through orifices may be obtained by knowing the time in which a cylinder, &c. is emptied by the water running out through a hole in the bottom.

Let $x =$ the depth, in feet.

$z =$ the area of the falling surface, in inches.

$n =$ the area of the stream, or orifice, in inches.

$s = 16$ feet, the space through which a body descends, by gravity, in one second.

$m = 32$ feet, the velocity acquired at the end of one second.

Then, As $\sqrt{s} : m :: \sqrt{x} : \frac{m\sqrt{x}}{\sqrt{s}}$, the velocity of the spout per second, at the first instant. And, As $z : n :: \frac{m\sqrt{x}}{\sqrt{s}} : \frac{nm\sqrt{x}}{z\sqrt{s}}$, the velocity of the falling surface per second. Also, As $\frac{nm\sqrt{x}}{z\sqrt{s}}$: $1'' :: \dot{x} : \frac{z\sqrt{s}\dot{x}}{nm\sqrt{x}}$, the fluxion of the time of emptying; the fluent of which is $\frac{2z}{nm}\sqrt{sx}$, $= \frac{z\sqrt{x}}{n\sqrt{s}}$.

Experiment 1. A cylindrical tube 3.166 feet long, and 2.705 inches in diameter, was exhausted in 3 minutes and 16 seconds, through a hole the area of which was .0141.

Experiment 2. The area of the cylinder and orifice the same, the length of the tube 6.458 feet, the time of exhausting, by observation, is 4 minutes and 40 seconds.

In Experiment 1, $\frac{z\sqrt{x}}{n\sqrt{s}} = \frac{10.23144}{.0564} = 3$ minutes and $1\frac{1}{2}$ second, which would be the true time, if at the depth of 16 feet the velocity was 32 feet per second. In Experiment 2, $\frac{z\sqrt{x}}{n\sqrt{s}} = \frac{14.605668}{.0564} = 4$ minutes and 19 seconds, the time of exhausting, by theory.—From these experiments, to find the velocity per second, when $m = 2s$. If $t =$ the time of exhausting, we have, from the above expression, $\frac{z\sqrt{x}}{n\sqrt{s}} = t$; from which we find $s = \overline{\frac{z\sqrt{x}}{tn}}^2$, and from the first experiment is equal to $\overline{\frac{10.23144}{2.7636}}^2 = 13.6974$ feet: in the second experiment, $s = \overline{\frac{14.605668}{3.948}}^2 = 13.6863$ feet.

In these experiments there is no difficulty in observing the time to half a second, or nearer. The length of the tubes can also be truly measured; but to obtain the area to a greater degree of accuracy, the water was weighed, reduced into inches, and the section computed. Taking the mean of these two experiments, at the depth of 13.6918 feet, the velocity is

27 3836 feet per second. And As $\sqrt{13.6918}$
: 27.3836 :: $\sqrt{16}$: 29.6 feet, the velocity at
16 feet, according to these experiments.

*Experiments on the Velocity with which Water
flows through syphons.*

FIG. XXIII.

AB is a vessel, &c. containing water, filled to
the line *ss*; *svb* a bended tube, or syphon, *sv*
the shorter, *vb* the longer leg; *ve* the perpendicular altitude of the crown of the syphon
above the level of the water to be conveyed
through it; *sb* the perpendicular height of the
difference of the two legs: the whole being
filled, this difference is the power which gives
motion to the whole; for if both ends were
placed in the same vessel of water, whatever
might be their lengths, their endeavour to descend would be as the perpendicular heights,
and these being equal, there could be no
motion. When *v* is not elevated more than
30 feet, the water in *vb* begins to descend, and
the weight of the atmosphere causes that in
the vessel AB to follow up the tube *sv*.

Perpendicular height of the syphon above the level of water.	Perpendicular length below the level.	Area of the orifice.	Weight per minute in ounces avoirdupoise.	Computed quantity per minute.	Square root of *sb*.
21 In.	24.	.0141	3.75	3.88	4.899
39 ..	27·5	....	4.075	4.158	5.244
58 ..	26.5	....	4.095	4.064	5.147
62 ..	55.75	...	5.819	5.904	7.466
62 ..	35.75	....	4.63	4.76	5.98
97 ..	9.25	....	2.245	2.41	3.041
95½..	20.	....	3.317	3.542	4.472

Without regard to the height of the top of the syphon above the water, the quantity or weight discharged is nearly as the square root of the height *sb*. For, though the weight of the water in the tube does for a moment retard the velocity, yet when once put in motion, that resistance, by these experiments, is not worth notice. And while the top keeps free from air, we may compute the velocity by taking the difference of the legs, and proceeding in the same manner as in finding the velocity through the bottom or side of a vessel.

Experiments on the Velocity with which Water ascends into a vacuum, by the pressure of the Atmosphere.

To make these experiments, I procured a number of tubes, of different lengths, the whole when screwed together was near 40 feet long; these were fixed in a stair-case, in such a manner that I could make use of the whole or of a part, as the experiment required; the highest was furnished with a stop cock at the top, so that when the water was raised by a small pump, by turning the cock, the tube below always remained full of water. Next, the receiver, Fig. XXV. was exhausted, accurately weighed, then screwed upon the pipe or tube, the cock opened, and the water suffered to flow into the vacuum for any convenient time, which was measured by a time-piece. The receiver was then unscrewed, and weighed again, by which the weight of the water which entered in a given time was known; and as the area of the orifice was also known, the velocity was easily computed.

Number.	Height at which the water is delivered.	Time, in seconds.	Weight of water raised in ounces avoirdupoise.	Cubic inches.	Area of the hole.	Velocity per second, in feet.
1	8 inches	43	26	44.92	.002356	36.95
2	8 ..	45	27.5	47.52	....	37.35
3	8 ..	60	36	62.2	....	36.66
4	10 ..	36	22	38	....	37.33
5	10 ..	44	27	46.65	....	37.42
6	4.5 feet	36	20	34.56	....	33.95
7	4.5 ..	47	27	45.65	....	34.3
8	7.4 ..	50	27	46.65	....	33.00
9	7.2 ..	49	26.4	45.62	....	32.9
10	7.2 ..	51	27.2	47	....	32.7
11	7.5 ..	60	32	55 29	....	32.59
12	13.8 ..	60	26.3	45.44	....	26.78
13	13.8 ..	60	26.5	45.79	....	26.99
14	18.2 ..	60	23.7	41	....	24.17
15	25 ..	60	16.5	28.51	....	16.8
16	25 ..	60	17	27.64	....	16.3
17	32 ..	60	6.76	11.68	....	6.89

When these experiments were made, the barometer stood from 29.5 to 30 inches, and as the pressure is frequently varying, the velocity of the rising water must also vary. At the surface of the water, under the greatest and least pressure, the velocities would be in

the ratio of 37.6 to 36.   At the altitude of 16
feet, as 30 to 27.   At the height of 26 feet,
as 23 to 17, &c.   In these experiments, the
area of the hole is small, compared with the
area of the lower part of the tube: if the tube
was equally wide, as in the lower part of a
pump, the velocity of the water would be
much less.   And when a column of water is
to be moved by the weight or spring of the
air, as in a pump, the water is not instanta-
neously put in motion; and the longer the
tube, the longer is the water in acquiring its
greatest velocity.

N. B.   In the above experiments, I frequently
made the vacuum without an air-pump.   Upon
the tall tube, above the cock, I screwed as much
more tube as made the whole height about 36
feet; then an open receiver, furnished with a
stop cock at the top, was screwed upon the
tube, the whole was filled with water, and co-
vered with leather and a plate, with the jet-
pipe and cock; after this, the cock in the tube
was opened, and the water descended out of
the receiver and down the pipe or tube, till
the pressure of the air could support it: the
cocks being turned, the receiver was unscrewed, perfectly exhausted.

R

A

Treatise on Mills.

PART IV.

EXPERIMENTS ON CIRCULAR MOTION.

In this part, many of the principal Theorems and Problems contained in the first and
second parts of this work are confirmed by
experiments.

The machine, Fig. XXIV, is useful for
making experiments on circular motion; and
also for comparing the effects produced by a
given weight in falling through a given space,
when applied at different distances from the
axis, &c. The cylinder A is 2 inches and B 4
inches diameter. The wheel w has 84 teeth,
and the pinion P has 8. The balls c c may be
of any weight, and fixed at any distance from
the centre G. In a rod or arm revolving round
a centre, it will be as the velocity of any point

in that arm is to its distance from the centre, so is the velocity of any other point to its distance, &c. The balls c c make $10\frac{1}{2}$ turns while the cylinder A makes 1; therefore, if the distance at which they are placed be multiplied by $10\frac{1}{2}$, the product will be the distance at which a body must be placed from the axis of the cylinder A; and to revolve with it, the velocity of which shall be the same as that of the ball c.

For example,

Let the balls be at the distance of 6 inches from c, this multiplied by $10\frac{1}{2}$, gives 63, at which distance a body revolving round with the cylinder A, would have the same velocity with the balls c c. But their central forces, being as the squares of the velocities divided by their distances, would be different: in this case, as $\frac{1}{6} : \frac{1}{63}$, for their velocities are equal, or $v = v$, but their forces are $\frac{v^2}{D}$ and $\frac{v^2}{d}$.

In the machine, the wire arms are each 8 inches long, and weigh 22 dwts. together; the distance of the centre of gyration is 4.61 inches from their centre, where the resistance is equal to their weight, but reduced to the end of the arm is only equal to 7.33 dwts.; and at any distance d the resistance will be $= \dfrac{\overline{4.61}|^2 \times 22}{d^2}$.

EXAMPLE.

Let $r =$ the radius of the cylinder from
           which the weight M is suspended.
    $s =$ the distance c c $\times$ 10.5.

Then will $\frac{Mr^2}{s^2}$ be the force exerted at c to
give motion to the balls.

Let $r = 1$.
     $s = 84$.
     M $= 8$ oz.
     c $+$ c $= 8$ oz.

Then will $\frac{Mr^2}{s^2} = .0011336$ oz. the moving
power; the resistance is the moving power
itself, ($= .0011336$) the arms reduced to the
end, ($= .3666$) and the balls, ($= 8$), the sum
is 8.3677996 oz., the whole resistance. If
the moving power is divided by this, we have
$\frac{.0011336}{8.3677996} = .0001356$, the accelerating force
compared with gravity, which multiplied by
16, gives .0021469 feet the first second. And
As $1''^2 : .0021469 :: 100^2 : 21.469$ feet, the
space through which the body would descend
in 100 seconds. By which process the following experiments may be compared with each
other, and with the theory.

For, let $w =$ the weight of the fly.

$t =$ the time of descent, in seconds.

$F =$ feet passed over by M in the time t.

$d = 16$ feet.

$n =$ the space passed over by the fly, in the time t.

Then will $n = \dfrac{Msr}{ws^2 + Mr^2} \times dt^2$, Theo. 1.

$F = \dfrac{Mr^2}{ws^2 + Mr^2} \times dt^2$, Theo. 2.

$w = \dfrac{Mdr^2t^2}{Fs^2} - \dfrac{Mr^2}{s^2}$,

or $w = \dfrac{Mrdt^2}{ns} - \dfrac{Mr^2}{s^2}$, $\Big\}$ Theo. 3.

$M = \dfrac{Fws^2}{dr^2t^2 - Fr^2}$, Theo 4.

$t = \sqrt{\dfrac{Fws^2}{Mdr^2} + \dfrac{F}{d}}$, Theo. 5.—See Prob. XV. Part II.

In these Theorems, the resistance of the air is not considered; but the following experiments will be sufficient to convince us that when the moving body is small, and the velocity great, it will make a difference in the conclusion, as it prolongs the time, &c. And although it is of little importance to consider its effects on heavy bodies which move slowly, yet in making experiments to prove the truth of a theory, it

becomes necessary to take notice of every im-
pediment. The weights screwed upon the
arms were 2.5 inches long, 1 broad, and a little
thicker at the middle than at the ends (see Fig.
XXVI.); these could be fixed so that either
the end or the flat side should strike directly
against the air.

Experiment 1. The fly being fixed to meet
with the least resistance, a weight of 5.7 oz. suspended from the cylinder A, descends through
9 feet in 62 seconds.

Experiment 2. The fly being placed in a ver-
tical position, so that the flat side might press
against the air, 15 oz. suspended from the cylinder A, descends through 9 feet in 62 seconds;
but when the falling weight has acquired a velocity of 2 inches per second, it ceases to be
accelerated, or moves with an uniform motion.

Experiment 3. The fly being placed or fixed
to meet with the least resistance, 7.3 oz. at the
cylinder A descends through 5 feet in 40 seconds, at the cylinder B through 5 feet in 20
seconds.

Experiment 4. The flat side being set to
move against the air, 7.3 oz. suspended from

B falls through 5 feet in 23.5 seconds, but suspended from A, in 55 seconds.

Observations.—1. When the velocity is the same, the resistance, so far as it arises from the friction and inertia of the machine, must be the same. 2. In experiments 1st and 2d, the velocities are the same, but the weights or powers which produce them are 5.7 and 15; hence their difference 9.3 is employed to overcome the resistance of the air. 3. In both experiments, when the motion is uniform, the velocity of the fly is 14 feet per second; and 9.3 oz. divided by 84, gives .11071 oz., the force exerted on the fly to overcome the resistance of the air. 4. The difference of surface opposed to the air in the two experiments is 4.7 inches, by which divide .11071 oz., and we have .0235 oz., the resistance against one inch; by computation it is .02123 oz. 5. The resistance arising from the air being as the square of the velocity, the 3d and 4th experiments will be found to agree with each other, and with theory, but will not if the resistance of the air is neglected.

In the following experiments, the fly is so fixed as to meet with the least possible resistance.

TABLE. I.

No.	Power. M.	Cylinder. r.	Fall, inches. F.	Time, sec. t.	Turns of the fly.
5	7.3	A	90	52	152
6	7.3	B	90	26	76
7	7.3	B	22½	13	19
8	7.3	A	22½	26	38
9	14.6	A	90	36½	152
10	14.6	A	45	26	76
11	29.2	A	90	26	152

The resistance of the machine is found by Theo. 3. If the fly is used, its weight must be subtracted frow w, and a weight is left which is equal to the whole resistance reduced to the place of w. But in these experiments the weight and magnitude of the fly remain the same.

1. It appears (No. 5 and 6), that by increasing the distance at which the power is applied, we diminish the time of descent, but the power in falling through a given space produces effects which are as the times of descent.

2. When different powers are applied to the same cylinder, as in the 5th, 9th, and 11th experiments, they produce effects which are directly as the square roots of these powers, and inversely as the times. For, As $\sqrt{7.3} : \sqrt{14.6}$ :: $36\frac{1}{2} : 52$, nearly; and As $\sqrt{7.3} : \sqrt{29.2}$ :: $26 : 52$. The effects are the same, and the times are inversely as the square roots of the

powers, or the powers are inversely as the squares of the times, or directly as the squares of the velocities.

3. From the 5th and 10th it appears, that if a given weight in falling through a given space produces a certain effect, double that weight in falling through half the space produces half the effect.

4. In the 11th experiment the velocity is twice as great as in the 5th, but four times the power is employed to produce it: that is, the power is as the square of the velocity.

5. When equal effects are produced in the same time, the powers will be inversely as the distances at which they are applied, but the spaces descended through will be directly as the said distances. See experiments 6th and 10th.

TABLE II.

No.	Power, in ounces.	Cylinder.	Weight of the fly, in ounces.	Radius of the fly, in inches.	Fall, in feet.	Time, — in seconds.	Force acting on the fly, in ounces.	Turns of the fly.	Space passed over by the fly.	Weight of the fly, multiplied by the space.	Relative velocity of the fly.	Ceutrifugal force of the fly.
12	4	B	4	8	6	26	.0952	61	254	1016	2	2
13	4	B	16	4	6	26	.1904	61	127	2032	1	4
14	4	B	8	5.65	6	26	.1346	61	179.8	1438.7	1.414	2.8
15	16	A	4	8	6	26	.1904	122	508	2032	4	8
16	4	A	4	4	6	26	.0952	122	254	1016	2	4
17	1	B	4	4	6	26	.0476	61	127	508	1	1
18	2	A	4	4	3	26	.0476	61	127	254	1	1

S

In making these experiments, when the cylinder B was used, there was an addition of 11 dwts., and when A was used of 44 dwts., to the power; which weights alone would turn the machine, and descend through 6 feet in 26 seconds: so that the power in the Table was wholly exerted in turning the fly.

The numbers under *Force acting on the fly*, are obtained by multiplying the radius of the cylinder used, by the power, and dividing by the radius of the fly multiplied 10.5. For example, No. 12, the radius of B is 2 inches, which multiplied by the power, 4 oz., gives 8, which divided by 84 (10.5 × 8), the radius of the fly, the quotient is .0952, and so for the rest.

The products of the weight of the fly into the space it has described, are always as the force acting on the fly: that is, when the power employed is, by the above process, reduced to the fly, the effects are as these powers, whether the fly be heavy or light, or whether the circle it describes be large or small.

The space passed over by the fly is obtained by multiplying the circumference of the circle in which it moves, by the number of turns.

The relative velocity is the space reduced to less numbers by division.

The centrifugal forces are computed by the Theorems in the first part; but here the numbers are not any real measure, but comparative; for as the velocity is constantly increasing, so is the central force. At the end of the time, the true centrifugal force, in experiment 12th, is nearly 17 times the weight of the revolving body, or 68 oz.; and the true force, in ounces, at the end of the other experiments, will be had by multiplying that in the Table by 34.

When the cylinder B is used, the angular velocity is always the same; as it is also when A is used (except No. 18), but is twice as great as when B is used, or the revolutions are performed in half the time.

When the power, and distance at which it is applied, increase with the radius of the fly, the angular velocity will remain the same, No. 12, and 18; for 3 feet of cord uncoiled from A will produce as many revolutions as 6 feet from B.

The velocities are as the square roots of the powers which produce them, or the powers are as the squares of the velocities, No. 12 and 17.

When any number of revolving bodies, each multiplied by the square of its distance from the centre, have their products equal, by a given power, equal angular velocities will be produced, No. 12, 13, 14, 17.

If a given power communicates a certain velocity to one quantity of matter twice, thrice, &c. that power will communicate the same velocity to twice, thrice, &c. that quantity of matter, No. 13 and 17.

If the weight of a revolving body, and time of a revolution, remain the same, the power will be as the square of the distance, No. 12 and 17.

The centrifugal forces are as the impelling powers, when these powers are applied in the same manner, No. 13 and 17; for the radius of the circle being the same, the forces are as the squares of the velocities, or as the powers which produce them.   See Prob. XIII, XIV, &c. Part II.

Experiments on Water Wheels.

FIG. XXVII.

A B C D is a water wheel with buckets, the circumference is 5.6 feet, or the diameter of the circle passing through the centre of the buckets is 20 inches. By turning a screw, this wheel may be taken off the axle, and one of a different size put in its place. The cog wheels may also be changed at pleasure; as also the wheels or trundles on the second axle, which are turned by the cog wheels. On the second axle is another cog wheel, with 84 teeth; this turns an upright spindle, with 8 teeth or leaves; on the top of this spindle is fixed a fly, the weight and radius of which may be varied at pleasure.

FIG. XXVIII.

The plate F F is divided into 60 parts, and the index I makes one revolution while the water wheel makes 60. The index L is fixed on the axle of the water wheel, and therefore makes the same number of revolutions: the plate G G, over which it moves, is divided into 10 parts. There is also another index, not visible in the figure, which makes one revolution while the water wheel makes 600.—By these, and a time-piece, to measure the minutes and seconds, &c. the number of revolutions made by the water wheel in any given time may easily be known.

The water is conveyed from the cock N into the trough H, or into the vessel F G E, which has three apertures or sluices at 1, 2, 3; these apertures are so adjusted by trial, that when the vessel is filled to F G, if any one is opened, it will discharge the water exactly as fast as it is

supplied by the cock; so that the water acts
upon the wheel both by weight and impulse.
The cistern which supplies the cock is always
kept equally full; that is, the water is returned
by a pump, and the person who works it takes
care to keep the surface always at the same mark.

ON THE APPLICATION OF WATER.

On this subject there have been various opinions; but opinions unsupported by theory or
experiment claim little regard.   At whatever
altitude the water might be applied to the wheel,
it has been the practice with many to diminish
that altitude, in order to obtain a head, as it has
been supposed that the impulse produced by
that head would have a greater effect than the
water by its gravity could produce, if thrown
into the buckets immediately from the surface.
And no doubt but it is owing to opinions of
this kind, that the water, which might be applied at the top, is often caused to strike, or
fall upon the wheel, about 45 degrees from the
vertex.   The following experiments, which
have been often repeated, both in public and
private, and with the greatest accuracy and attention, give no encouragement to diminish the
altitude for the sake of obtaining more advantage by the impulse.

Whether the effect is measured by the velocity, its square, or its cube, every thing remaining the same, it will be granted, that when the wheel moves quickest, the water is applied to the greatest advantage.

Experiment 1. The orifice No. 1, is opened, the water strikes the wheel near the bottom, and acts by impulse and gravity; the head is equal to the diameter of the wheel. The velocity of the stream is 6.5 feet per second; the water wheel makes 60 revolutions in 7 minutes and 21 seconds, or is turned 8.2 times in one minute.

Experiment 2. The sluice No. 2, is opened, the water strikes the wheel near the centre, and causes it to revolve 60 times in 3 minutes and 57 seconds, or 15.19 times in a minute.

Experiment 3. The orifice No. 2, is shut and No. 3 opened, the water falls upon the wheel, with impulse, at 45 degrees from the vertex. The wheel revolves 60 times in 3 minutes and 28.5 seconds, or 17.26 in a minute.

Experiment 4. The water is conveyed into the trough H, from which it falls upon the vertex of the wheel; 60 revolutions are made in 3 minutes and 15 seconds, or 18.46 in a minute.

In these experiments, the fly makes 21 revolutions for 1 of the water wheel, its weight
was 4 oz. and diameter 13 inches. The comparative velocities of the water wheel or fly,
will be as the number of turns made in a given
time, or as 8.2, 15.19, 17.26, and 18.46.

In the following Table we have the whole in
one view.

TABLE III.

Experiments.	Sluice.	Turns of the water wheel per minute.	Relative velocity.	Turns of the fly per minute.	True velocity of the fly, in feet, per second	True velocity of the water wheel, in feet, per second.	
1	No. 1	8.2	1	172.2	11.96	.765	
2	2	15.19	1.85	319	22.15	1.417	
3	3	17.26	2.1	362.5	25.17	1.61	
4	overshot	18.46	2.25	387.6	26.91	1.723	
5	No. 1	4.	1	168	11.66	.373	*In these four experiments, the fly revolves 42 times for the water wheel once; its radius 6 inches.*
6	2	7.69	1.92	323	22.43	.717	
7	3	8.88	2.22	373.3	25.92	.829	
8	overshot	9.52	2.38	400	27.77	.888	
9	No. 1	3.05	1	128.1	8.89	.284	In these, the radius of the fly is 8 inches, weight 12 ounces; revolutions 42 to 1.
10	2	6.38	2.09	268	18.61	.595	
11	3	7.28	2.38	305.7	21.23	.679	
12	overshot	7.72	2.53	324.2	22.51	.720	
13	No. 1	2.14	1	134.8	9.36	.109	The radius as above; weight 4 oz.; revolutions 63 to 1.
14	2	5.	2.33	315	21.87	.466	
15	3	5.84	2.72	368	25.55	.545	
16	overshot	6.4	3	403 4	28 01	.597	

In the above four sets of experiments, which
are considerably varied, a given quantity of
water produces the greatest number of revolutions when applied at the top of the wheel.

The ratio of the revolutions when the water is applied at the top, is to the revolutions when applied at the quarter, nearly as 32 to 30, or as 16 to 15.   For, in the first set, As 2.25 : 2.1 : : 16 : 14.9.   In the second set, As 2.38 : 2.22 : : 16 : 14.9.   In the third set, As 2.53 : 2.38 : : 16 : 14.05.   In the fourth set, As 3 : 2.72 : : 16. : 14.5.

From these experiments it appears, that for every 15 revolutions when the water is applied at the quarter, there will be 16 when the water is applied at the top of the wheel.

Experiments on a wheel 10 inches diameter. The water applied at the top produces 60 turns in 4 minutes and 25 seconds, but when applied to the quarter, 60 turns in 4 minutes and 52 seconds; the turns per minute in the first case are 13.58, in the second 12.32; and As 13.58 : 12.32 :: 16 : 14.51, the number of revolutions, when the water is applied at the quarter, for 16 when it is applied at the top, or a given quantity of water in the overshot produces 16 turns of the water wheel in the same time as it produces 14½ when applied at the quarter.

Three quarts of water in a minute, laid on at the top of the wheel, turns it 60 times in 6

T

minutes and 45 seconds; four quarts set on at
the quarter produces the same number of revolutions in the same time. And if the weight
of the fly or the ratio of its revolutions to those
of the water wheel be varied, yet the ratio of
the quantities of water to produce the same
effects in the same time remains the same.

Experiments on the Velocity of Wheels.

In these experiments, the effect is measured
or estimated by the number of the revolutions
which a fly makes in a given time: and although
this is not the true measure of the power of
the water, (part of it is exerted in turning the
wheels), yet it is as much so as the revolutions
of a mill stone are the measure of the effect
produced by the water; and that velocity which
a wheel has when a mill stone makes the greatest number of revolutions in a given time, will
certainly be considered as the best. It may be
observed, that these experiments are confined
to wheels on which the water acts by gravity
only, and that the buckets are supposed to be
large enough to retain the water till it comes
near the bottom of the wheel: for if the buckets lose part of the water as they descend, the
wheel, in its present construction, is not doing

the most work.—In the following experiments the quantity of water was always the same; the water wheel made 60 revolutions, and sometimes more; the time was observed, from which the number made in one minute was computed.

In order to vary the velocity of the water wheel, it becomes necessary that the ratio of its revolutions, to those of the fly, should also vary, which is effected by changing some of the wheels.

Experiment 1. The water is applied at the top in this, and all the following experiments under this head. The wheel makes 15.36 revolutions in a minute, the fly 21 for 1; and 15.36 × 21 = 322.56, the number of revolutions which the fly makes in a minute.

Experiment 2. The water wheel turns 12.08 times in a minute, and the fly 28 for 1; hence the fly revolves 338.24 times per minute.

Experiment 3. The fly turns 44.6 times for the water wheel once, which is turned 7.67 times in a minute; of consequence, the fly revolves 342.08 times per minute.

Experiment 4. The fly makes 59.5 turns for the water wheel once, and the water wheel turns 6.02 times in a minute; the fly, therefore, turns 358.19 times per minute.

But perhaps these four experiments, with others made with a different fly, may appear to more advantage as classed in the following Table.

TABLE IV.

Experiments.	Turns of the water wheel per minute.	Velocity of the water wheel, per second, in feet.	Turns of the fly for one of the water wheel.	Turns of the fly per minute.	Ratio of the quantity of water in the buckets.	Quantity per minute to turn the fly equally fast.	Turns of the water wheel per minute, when the fly turns equally fast.	Quantity on the wheel at once, when the fly has the same velocity.	Velocity of the water wheel in feet, when the motion of the fly is equable.
1.	2.	3.	4.	5.	6.	7.	8.	9.	10.
1	15.36	1.433	21	322.5					
2	12.08	1.127	28	338.2					
3	7.67	.716	44.6	342					
4	6.02	.562	59.5	358.2					
5	36	3.36	8.75	315	1	4	36	1	3.36
6	28.125	2.62	13⅛	369.1	1.282	3.11	24	1.166	2.24
7	18.18	1.69	21	381.8	1.988	2.45	15	1.47	1.4
8	15.58	1.45	25	389.5	2.317	2.29		1.606	1.666
9	12.57	1.11	31½	395.6	2.863	2.14	10	1.926	.933
10	9.52	.88	42	400	3.812	2	7.4	2.4	.7
11	7.67	.71	52½	402.9	4.732	1.90	6	2.8	.56
12	6.4	.59	63	403.4	5.7	1.82	5	3.368	.466

The second column shews the turns of the water wheel per minute, as found by experiment.

The third is found from the second: for example, in the first experiment, 15.36 multiplied by 5.6 feet, the circumference of the wheel, gives 86.016, the feet passed over in a minute, which divided by 60, gives the feet per second, = 1.433.

The fourth column is known from the wheels made use of.

The fifth is the product of the second multiplied by the fourth.

The sixth shews the ratio of the quantity of water hanging upon the wheel at once, which is always inversely as the velocity of the wheel: for instance, the wheel moving with a certain degree of velocity, every bucket receives a given quantity, but if the wheel moves twice as fast, a bucket can receive but half the quantity; or if the wheel has half the first supposed velocity, a bucket will receive a double quantity. Otherwise, suppose the wheel moves 6 feet in a second, and that a bucket receives 10 gallons; then, if it moves but 3 feet per second, a bucket will receive 20 gallons, or if 2 feet per second, 30 gallons, &c. In the fifth experiment, the wheel turns 36 times in a minute; in the ninth, but 12.57 times: hence, inversely, As 36 : 1 portion of water : : 12.57 : 2.863 times the quantity.

By increasing the velocity of the fly, compared with that of the water wheel, we have, in the last experiment, near six times as much water upon the wheel at once, from a given stream, as there is in the fifth experiment, in which the fly is turned 315 times in a minute; but in the last, 403 times.

In the seventh column we have the quantity
of water as found by experiments to produce
315 revolutions of the fly in one minute; and
it appears, that by increasing the velocity of
the fly compared with that of the water wheel,
that a less quantity of water is sufficient to
produce the same effect, or the same number of
revolutions of the fly: for in the last experiment the quantity of water is not half of that
used in the fifth, though the effects are the
same.

The eighth column contains the number of
revolutions which the water wheel makes in a
minute, under the different changes of wheels
in the communication of motion to the fly,
when the fly itself makes the same number of
revolutions per minute.

The ninth column exhibits the comparative
quantity of water hanging on the wheel at once,
when the fly moves equally fast.

The tenth is the velocity of the water wheel,
in feet, when the fly has the same velocity in
all the experiments.

The five following experiments were made
with a wheel of 11 inches diameter.

TABLE. V.

Experiments.	Turns of the water wheel per minute.	Velocity of the water wheel, per second, in feet.	Turns of the fly for one of the water wheel.	Revolutions of the fly per minute.
13	14	.677	21	296
14	9.6	.464	31½	302.4
15	7.31	.353	42	307
16	5.9	.285	52½	309.75
17	4.94	.238	63	311.55

The speed of the water wheel being diminished, the revolutions made by the fly in a
given time are increased in number, both with
the large and small wheel. It is true they are
not increased in the same ratio when the velocity is much diminished, as in the eleventh and
twelfth in Table first, and again in the sixteenth
and seventeenth in Table second, but this is
owing to the buckets being overloaded, so that
they lost a part of their water too soon, and of
consequence the effect in the two last experiments is not much increased. But if the buckets retain their water till they come near the
bottom, it appears, that the slower the water
wheel moves, the greater its effect will be, provided the motion continues to be equable. See
Prob. XXXVIII, Part II.

This may also be proved by raising weight by
a cord coiled round the axle of the water wheel.

The following Table contains seven experiments, in which the velocity of the water wheel is gradually diminished from the first to the last; but the effect in the last is the greatest.

TABLE VI.

Experiments.	Weight.	Time of rising through 6 feet.	Velocity, or feet per second.	Effect produced.
1	$\frac{1}{2}$ *lb.*	57.5″	.1043	5.21
2	1	62.4	.0961	9.61
3	1$\frac{1}{2}$	69.5	.0863	12.94
4	2	75.5	.0795	15.90
5	3	93	.0645	19.35
6	4	117	.0512	20.48
7	5	142	.0422	21.10

To find the velocity per second. As 57.5″: 6 feet : : 1″ : .1043 feet; and As 62.4″ : 6 feet : : 1″ : .0961 feet, and so for the others. The effect is obtained by multiplying the velocity thus found by the weight raised, as in the first experiment, $\frac{1}{2}$ lb. multiplied by .1043, gives .0521, the effect; and in the second, 1 lb. multiplied by .0961 = .0961, and in the same manner the effect in all the other experiments is obtained. In the Table, the effect thus found is multiplied by 100, or the decimal point is removed two figures, but the ratio is the same.

If it was required to find the weight raised in a given time, suppose 1 hour, it would be, *As the time of rising through 6 feet is to the weight raised so is 1 hour to the number of times which the said weight would be raised through the given space.*

For example, As 57.5″ : ½ lb. : : 3600″ (1 hour) : 31.304 pounds per hour. In the third experiment, As 69.5″ : 1.5 lb. : : 3600″ : 77.7, pounds per hour. In the sixth experiment, As 117″ : 4 lb. : : 3600″ : 123.08, pounds per hour. In the seventh experiment, As 142″ : 5 lb. : : 3600″ : 126.7, pounds per hour.

Which experiments so far, in conjunction with those made with the fly, prove that by diminishing the velocity of the water wheel, we increase the power.

On the Velocity communicated to a Wheel, by different Quantities of Water.

See page 17,—where it appears that if a given stream produces a certain velocity in the wheel, a stream eight times as large will be required to communicate double that velocity, or to turn the wheel twice round in the same time. And for any other quantity of water which may be applied, it will be, *As the cube root of the present quantity is to one degree of velocity, so is the cube root of the other quantity to the velocity it will produce.*

v

The truth of the proposition will appear by the following experiments.

FIG. XXX.

Pieces of brass were screwed upon the end of the cock, and perforated, by trial, till the aperture was large enough to deliver the exact quantity required: for instance, a pint, a quart, two quarts, &c. per minute, and which quantity was always applied in the same manner.

Experiment 1. One pint of water per minute turns the water wheel 60 times in 13 minutes and 33 seconds.

Experiment 2. Two pints per minute turn it 60 times in 10 minutes and 42 seconds.

Experiment 3. Four pints per minute produces 60 turns in 8 minutes and 27 seconds.

Experiment 4. Six pints per minute produces 60 turns in 7 minutes and 18 seconds.

Experiment 5. Eight pints per minute produces 60 turns in 6 minutes and 18 seconds.

In the following Table we have, in the first column, the quantity of water applied; in the second, the cube root of that quantity; in the third, the comparative velocities as found by the experiments; in the fourth, the number of turns the water wheel made in a minute.

TABLE VII.

Experiments.	Water per minute.	Cube root of the quantity of water.	Velocity of the wheel.	Turns per minute.
1	1	1	1	4.42
2	2	1.2599	1.267	5.6
3	4	1.5874	1.606	7.1
4	6	1.8171	1.859	8.219
5	8	2	2.15	9.52

The number of turns per minute is obtained by direct proportion: for, in experiment 1, As $13'\ 33''$: 60 revolutions :: $60''$: 4.42 revolutions, and so for the rest.

The velocity is directly as the number of turns made in a given time; and if in the first experiment it is represented by 1, in the second it will be 1.267; for As 4.42 : 1 :: 5.6 : 1.267, and As 4.42 : 1 :: 7.1 : 1.606, &c.

In the fifth experiment, the quantity per minute is eight times as much as in the first; but it may be observed, that the load or weight upon the wheel is only four times as much as in the first; for as the wheel moves twice as fast, a bucket can only receive half as much water as it would do with the velocity it had in the first experiment.

ON THE SIZE OF WHEELS.

THE fall being given, or the greatest altitude at which a given stream can be applied to a wheel, it becomes necessary, both on account of expence and utility, to enquire what the diameter of the wheel ought to be.—This may be determined either by theory or experiment.

1. Let us suppose two wheels equally heavy, the diameter of one 10, and the other 20 feet, and that the weight is chiefly in the circumference, or as much so in one as in the other.

2. Let the same power be applied to the circumference of each wheel, and each circumference will have the same velocity; hence, the time of a revolution will be as the diameters directly, or in the present case as 1 to 2, or the 10 feet wheel will be turned twice for the 20 feet once, as already demonstrated in Prob. XIV, Part II, and proved by experiment in No. 12 and 16 in Table II, page 129. In both experiments the power is 4 oz., and the weight of the fly or wheel is the same; but in No. 12 the radius is 8 inches, and the power is applied at the distance of 2 inches; in No. 16 its radius is 4 inches, and the power is applied at 1 inch: hence the ratio of the effects produced

is exactly the same as if it had been applied to the circumference of the two wheels; and as their weights are equal, and they move equally fast, the effects are also equal: but their revolutions are not performed in the same time; for, in 26 seconds, one revolves 61 times, and the other 122, or the revolutions are performed in times which are as 1 to 2, as above.

3. The power being employed to turn a single wheel, the velocity of its circumference will vary with the weight of the wheel, and of consequence the time in which it revolves: for if one of the above wheels is made twice as heavy, it will be twice as long in making one revolution as it was before.

4. If the power is a stream of water, this will vary with the velocity of the wheel. Suppose a given stream to turn the wheel with two degrees of velocity, then if the velocity of the wheel is but one, there will be twice as much water upon it, or in the buckets, as when it had two degrees of velocity: hence, if a wheel of a given weight has a velocity of 4 feet per second, one 4 times as heavy would be caused to move 2 feet per second by the same stream (see page 113, and 125), not considering the water as part of the mass to be moved.

5. Let us suppose a stream of water applied at the top of a wheel 10 feet in diameter, and that it turns the wheel in 6 seconds; let the same stream be applied at the centre of another wheel 20 feet in diameter; then it is evident, that if every bucket receives as much water as before, that the large wheel will be twice as long in making one revolution as the small wheel, and therefore by the same cog wheel, &c. cannot do the same work; and if we encrease the cog wheel in the same ratio as the water wheel, we obtain no power by making the wheel larger.

The truth of the above is confirmed by the following experiments.

Experiment 1. A given stream is applied exactly at the centre of a wheel 20 inches in diameter; it makes 60 revolutions in 4 minutes and 21 seconds.

Experiment 2. The large wheel is removed, (the cog wheel, fly, &c. remain the same) and one of 10 inches in diameter is fixed in its place; the same quantity of water is applied at the top; and it makes 60 revolutions in 4 minutes and 15 seconds.

Experiment 3. The water is applied at the centre of the great wheel, and in addition to the fly, one pound is suspended from the axle, which is raised through 5 feet in 59½ seconds. The large wheel being removed, and the small one fixed in its place, every thing else remaining the same, the same weight is raised through the same space in 55 seconds.

In the last experiment, if the great wheel produces the same effect as the small wheel in the same time, the circumference must move twice as fast, by which the power would be diminished. See page 138, &c.

Experiment 4. A cylinder is fixed upon the axle, which has the same ratio to the diameter of the large wheel as the axle itself has to that of the small one, or the axle is 3 inches diameter, and the cylinder 6 inches; the water is applied at the centre of the great wheel, and the weight is raised in 56 seconds.

In the above experiments, except the third, there ought to be no difference in the effects produced by the two wheels but what arises from the difference of the weight of the said wheels and their velocities; hence the greater effects produced by the small wheel is owing to its requiring a less force to turn it, independent of the rest of the machinery, than what is necessary to turn the large wheel, or to cause it to revolve in the same time (see Prob. XIV, Part II.). For if the wheels were of the same weight, and the same power applied to the circumference of each, the small wheel would make two revolutions for the large wheel one.

ON THE FALL OF WATER.

By the *fall*, is understood the perpendicular altitude, measured from the bottom of the wheel to the surface of the dam, which, when the water issues upon the wheel at some distance below the surface, is frequently distinguished by *head* and *fall*.

The effect produced by a given stream in falling through a given space, if compared with a weight, will be directly as that space; but if we measure it by the velocity communicated to the wheel, it will be as the square root of the space descended through, agreeable to the laws of falling bodies.

Experiment 1. A given stream is applied to a wheel, at the centre; the revolutions per minute are 38.5.

Experiment 2. The same stream applied at

the top, turns the same wheel 57 times in a minute.

If in the first experiment the fall is called 1, in the second it will be 2. Then, As $\sqrt{1}$: $\sqrt{2}$ :: 38.5 : 54.4, which are in the same ratio as the square roots of the spaces fallen through, and near the observed velocity.

In the following experiments a fly is connected with the water wheel.

Experiment 3. The water is applied at the centre, the wheel revolves 13.03 times in one minute.

Experiment 4. The water is applied at the vertex of the wheel, and it revolves 18.2 times per minute.

And As 13.03 : 18.2 :: $\sqrt{1}$: $\sqrt{2}$, nearly.

From the above we infer, that the circumferences of wheels of different sizes may move with velocities which are as the square roots of their diameters without disadvantage, compared one with another, the water in all being applied at the top of the wheel. For the velocity of falling water at the bottom or end of the fall is as the time, or as the square root of the space fallen through: for example, let the fall be 4

x

feet; then, As $\sqrt{16} : 1'' :: \sqrt{4} : \frac{1}{2}''$, the time of falling through 4 feet. Again, let the fall be 9 feet; then, As $\sqrt{16} : 1'' :: \sqrt{9} : \frac{3}{4}''$, and so for any other space, as in the following Table, where it appears that water will fall through 1 foot in a quarter of a second, through 4 feet in half a second, through 9 feet in three quarters of a second, and through 16 feet in one second. And if a wheel 4 feet in diameter moved as fast as the water, it could not revolve in less than 1.5 seconds, neither could a wheel of 16 feet diameter revolve in less than 3 seconds; but though it is impossible for a wheel to move as fast as the stream which turns it, yet, if their velocities bear the same ratio to the time of the fall through their diameters, the wheel 16 feet in diameter may move twice as fast as that of 4 feet.

TABLE VIII.

Height of the fall, in feet.	Time of falling, in seconds.	Height of the fall, in feet.	Time of falling, in seconds.
1	.25	14	935
2	.352	16	1
3	.432	20	1.117
4	.5	24	1.22
5	.557	25	1.25
6	.612	30	1.37
7	.666	36	1.5
8	.706	40	1.58
9	.75	45	1.67
10	.79	50	1.76
12	.864		

When different streams of water produce equal effects, the quantities must be inversely as the fall. For instance, if 6 cubic feet in falling through 4 feet produces a certain effect, 3 cubic feet in falling through 8 feet would produce an equal effect; or the quantity being the same, the effect will be as the space fallen through.

The resistance afforded by mill-stones, and other kinds of machinery, cannot be computed from first principles. We can only ascertain it by observations: as for example,

In a Corn-mill,

Suppose 4 cubic feet per second, applied at the top of a wheel 12 feet high, will turn the mill-stone a given number of times per minute, what is the resistance of the whole, at the circumference of the water wheel?

Let the wheel revolve in 8 seconds; then will the buckets contain as much water at once as is delivered in 4 seconds, or half a revolution, viz. 16 cubic feet, or 1000lb. But as this weight is equally dispersed over one-half the circumference of the wheel, the weight acting at the extremity of the horizontal arm will be less; for the diameter being perpendicular, the centre of gravity of the circular arch will be

6366 of the radius, from the axis; therefore, from the principle of the lever, .6366 multiplied by the whole weight 1000lb., gives 636.6, the weight at the circumference, or at the end of the horizontal arm.

EXAMPLE II.

Suppose the wheel revolves in 10 seconds, every thing else remaining the same.

In this case, 20 cubic feet, or 1250lb. of water is contained in the buckets, which multiplied by .6366, gives 795.7lb. for the power at the end of the horizontal arm.

In the above examples I have supposed that the buckets contain their water till they come to the bottom; and if the wheel moves from 3 to 4 feet per second, the buckets may be made of such a form as to carry the water quite to the bottom, provided they are large enough, when at the centre, to hold near twice as much water as is necessary to do the work; and if the wheel moves faster, they will carry the water too far.

The Effects of Impulse and Gravity compared.

From the experiments contained in the Table page 115, it appears, that the impulsive force of a stream, is to the absolute weight of the column which impels it, as 16 to 10, or as 8 to 5, nearly; from which we may easily compare the impulse with the gravity of a given quantity.

For example :

Let the head be 4 feet, and the area of the sluice 1 inch; then will the velocity be 10.8 feet per second: that is, the square root of the depth multiplied by 5.4.

The solid content of the column which impels the stream is 4 feet multiplied by 1 inch = 48 cubic inches, and its weight is 27.77 oz. The quantity per second is 10.8 feet long, and 1 inch area, = 129.6 cubic inches, or 75 oz.

The velocity of the wheel is one-third the velocity of the stream, = 3.6 feet per second (see page 30). And as 5 : 8 :: 27.77 oz., the whole weight of the column, : 44.44 oz., the impulsive force upon an object at rest. But, the quantity being the same, the force is as the square of the velocity with which it strikes the wheel, in this case, equal to the difference between 10 8, the velocity of the stream, and 3.6, the velocity of the wheel, that is, 7.2 feet per second. Hence, As $\overline{10.8}^2 : \overline{7.2}^2 :: 44.44 :$ 19.6 oz., the impelling force constantly acting upon the wheel. Or if the velocity of the wheel should have any other ratio to the velocity of the stream, suppose 1 : 2, or that the wheel moves half as fast as the stream ; then, As $\overline{10.8}\rvert^2 : \overline{5.4}\rvert^2 : : 44.44$ oz. : 11.11 oz., the force acting upon the wheel.

In the Overshot.

Let the same quantity of water be thrown upon the top of a bucket-wheel 4 feet diameter.

The quantity per second is 75 oz. If the wheel moves as fast as before, which we will suppose for the sake of the contrast, it will revolve in 3.48 seconds, and will at once be loaded with as much water as is delivered in 1.74 seconds, viz. in the time of half a revolution, therefore 75 oz. × 1.74 = 130.5 oz., the weight constantly acting on the overshot wheel; but as it is not all gravitating at the end of the horizontal arm, but dispersed through a semi-circle, we must, as before observed, multiply the whole weight 130 5 oz., by .6366, the distance of the centre of gravity in the semicircle from the centre of the circle, and it gives 83 oz., equal to the effect of all the water at the end of the horizontal arm, which is more than four times the impulse of the same quantity.

EXAMPLE II.

Let the head be 9 feet, and area of the sluice 1 inch.

Then, the square root of 9 × 5.4, gives the velocity per second, = 16.2 feet; and 16.2 × 12, gives the cubic inches per second, = 194.4, which multiplied by .5787 oz., the weight of 1 inch is = 112.5 oz.; weight of a column 9 feet long and 1 inch square = 62.3 oz. Then, As 5 : 8 :: 62.3 oz. : 99.68 oz., the impulse upon an object at rest.

The velocity of the wheel $= \frac{16.2}{3} = 5.4$ feet per second, which taken from the velocity of the stream, leaves 10.8, the velocity with which the stream strikes the wheel. Then As $\overline{16.2}|^2$ $: \overline{10.8}|^2 :: 99.68$ oz. : 44 3 oz., the whole force of the stream upon the wheel.

In an Overshot.

The circumference of the wheel divided by the velocity, gives $(\frac{28.27}{5.4})$ 5.2 seconds for the time of a revolution, half of which multiplied by the weight per second, gives 292.5 oz. for the weight of water upon the wheel, when it moves with the above velocity, which being multiplied by .6366, gives 186.2 oz., the force exerted in the direction of a tangent to turn the wheel.

Many have expressed their surprise that the impulse of a given stream should be less than its gravity: here I have admitted the impulse to be almost double the gravity, but the reason why its effect is less does not seem very surprising.

OBSERVATIONS ON BUCKETS AND FLOATS.

IF the floats of a wheel are fitted to move in a channel of the same curvature (provided there was no loss of water), the effect of the water would be the same as in a bucket-wheel.

Demonstration.

FIG. XXIX.

Let c b b be part of a water wheel; and let the weight of the water in the bucket $a\,b\,d =$ w; the sine of the angle B c $d =$ c A $= s$; c B, the radius, $= r$. Then will $\frac{ws}{r}$ be the force exerted at B to turn the wheel in the direction of a tangent.

Next, if B $b\,d$ B represent a curved channel in which a float-wheel moves, then the water $a\,b\,d$ will be supported by the channel $d\,b$ and float $a\,d$. The sine of the angle which the channel $d\,b$ makes with the horizon will be equal to that of the angle B c d, and the pressure which the float sustains will be obtained from the property of the inclined plane, viz. *As the radius is to the sine of the angle of elevation, so is the*

whole weight to that part which the float sup-
ports; or, As $r : s :: w : \frac{ws}{r}$, the same as in the
bucket-wheel.

EXAMPLE I.

Let a bucket containing 100lb. of water be
at the distance of 30 degrees from the bottom,
to determine the force with which it acts on the
wheel, in the direction of a tangent?

The natural sine of 30° is .5, the radius
being 1 ; therefore, $\frac{100 \times .5}{1} = 50$lb., the whole
force of the water whether confined in a bucket
or in the channel.

EXAMPLE II.

Suppose the bucket or float at the distance of
50 degrees from the perpendicular?

The natural sine of 50° is .766, which multiplied by 100, the weight of the water, gives
76.6lb. for the force; and so for any other.

From which it appears that there is no difference, except the loss of water in the float-
wheel, for, however well it may be executed,
some water will pass; nevertheless, when the
stream is large, a wheel with floats may be found
more convenient than one with buckets. For
example, suppose a stream flowing over a dam
to be 6 feet wide, and 1 foot deep; the quan-

tity per second would be 21 cubic feet (see page 109). Suppose 3 buckets or floats pass the stream in 1 second, there will be 7 cubic feet for every bucket or float In a float-wheel there ought not to be more than six-eighths of the space filled with water ; and in a bucket wheel little more than one-half. The buckets, therefore, ought to contain about 14 cubic feet each, when at the end of the horizontal arm; and the space between the floats ought to be large enough to contain about 10 cubic feet; in which case, if the floats are 15 inches distant, they ought to be 16 inches deep. But for the bucket-wheel, the shrouds would be near 2 feet deep if large enough to contain the water. In which case, a float-wheel ought to be preferred ; and if a wheel is occasionally in back-water, floats are not so much retarded as buckets.

Experiments with Wheels acted on by Impulse only.

In addition to the experiments made on bucket-wheels, in which the water produces the effect by its *gravity*, I shall add a few made with float-wheels, on which the water acts by *impulse* only. If the wheel is confined in a channel, after the water has struck the float, it is retarded, and heaped against the float; in which case, the water is acting by impulse and gravity combined. In order to ascertain, as near as

possible, the true effect of the impulse, the following experiments were made. The theory
(page 30) assigns no more than a comparative
force to the water; or, if a given quantity issues
with a given velocity, the effect produced
should be as the square of the velocity with
which it strikes the wheel, and will of consequence vary with the velocity of the wheel;
and the velocity of the wheel will also vary
with the resistance. The effect of the stream
will be greatest when it strikes the floats in a
perpendicular direction, and will diminish as
the obliquity increases; and as the greatest force
is exerted on the first float, it was necessary, in
these experiments, to fix the floats a little inclined to the circumference of the wheel, or
instead of the common direction a s, (Fig.
XXXI) to direct them to the circumference of
a circle b b, in the direction b s, so that the direction of the stream may be nearly perpendicular to the first float s b, which is acted upon
till another intercepts the stream, during which
time its position is constantly changing.

The velocity of the stream was 9.3 feet per
second. The weight of water per minute was
10lb. The diameter of one wheel is 12 inches;
but I take the circumference of that part of the
floats which was struck by the water to be 2.8

feet, and the circumference of the other at 4.4 feet. The axle was truly turned, and supported on friction wheels, and by 41 revolutions raised a weight through 5 feet.

Experiments with the 12 inch Wheel.

TABLE IX.

Experiments.	Weight raised, in ounces.	Time of rising, 5 feet.	Velocity of the Weight, in feet, per minute.	Velocity of the wheel.	Velocity with which the wheel is struck.	Effect, by experiment.	Effect, computed.	Ratio of experimental effects.
	1.	2.	3.	4.	5.	6.	7.	8.
1	7	30	10	3.83	5.47	70	113.6	114.3
2	8	33	9.09	3.45	5.85	72.72	118.05	118.7
3	$8\frac{1}{4}$	35	8.57	3.3	6	72.8	118.8	118.9
4	9	37	8.11	3.1	6.2	72.9	119.16	119.16
5	$9\frac{1}{2}$	$39\frac{1}{7}$	7.6	2.91	6.39	72.2	118.76	118
6	10	46	6.52	2.5	6.8	65.2	115.5	106.5
7	11	56	5.35	2.05	7.25	58.8	107.14	96.1

Experiments with the 18 inch Wheel.

TABLE X.

1	9	$41\frac{1}{2}$	7.22	4.34	4.96	64.98	106.61	105.9
2	$10\frac{1}{2}$	$45\frac{1}{2}$	6.6	3.95	5.35	69.3	113.01	112.77
3	11	47	6.38	3.83	5.47	70.18	114.47	114.4
4	$12\frac{1}{2}$	52	5.77	3.47	5.83	72.12	117.9	117.5
5	13	54	5.55	3.34	5.96	72.22	118.6	117.65
6	$14\frac{1}{2}$	66	4.54	2.73	6.57	65.83	117.79	107.3

While the 12 inch wheel makes 41 revolutions, the part of the float which is struck passes over 115 feet, and the same part of the 18 inch wheel moves 180.4 feet: hence the velocity per second is found as follows: in experiment 1, As $30''$: 115 feet :: $1''$: 3.83 feet, &c.

The velocity with which the stream strikes the wheel is also known, by subtracting the velocity of the wheel from that of the stream; as in the first experiment, 3.83 taken from 9.3, leaves 5.47 for the velocity with which the wheel is struck.

Also, As 30″ : 5 feet : : 60″ : 10 feet, the space passed over by the weight in one minute; which space multiplied by the weight, gives the effect contained in the sixth column.

The greatest effect produced by the 12 inch wheel is in the fourth experiment, where 9 ounces is raised through 8.11 feet in a minute; their product is 72.9 for the maximum.

But the maximum in the 18 inch wheel is in the fifth experiment, where 13 ounces is raised 5.55 feet in a minute, and their product is 72.2, which is not quite so much as that of the less wheel; and, indeed, the difference would have been greater if the wheels had been equally heavy, but the larger wheel weighs 12 oz. and the smaller 1lb. 14 oz. It may also be observed, that the larger wheel is in this experiment rather short of the maximum, its velocity is 3.34 instead of 3.1, and in the next experiment it moves too slow, or is decreasing in power.

To compare the power with the effect.—We have 10lb. or 160 oz. of water falling through 2.2 feet in one minute; their product, 352, is the power; the effect is 73, or $\frac{73}{352}$ of the power $= \frac{206}{1000}$. But in the overshot (see page 144) the power is to the effect as 18 : 12.675, or as 1 : .704, or the effect is $\frac{704}{1000}$ of the power, which is 3.4 times as much as by impulse, though in these experiments we have the advantage of friction-wheels, which were wanting in the experiments with the overshot; besides, the overshot was encumbered with a fly, without which the motion would not have been regular.

We may further observe, that the effects obtained by multiplying the square of the velocity with which the water strikes the wheel, by the velocity of the wheel, as contained in the seventh column of the Table, agree, as near as can be expected, with the experiments, till we come at a maximum; after which, the effect, as found by experiment, varies from that computed by the square of the velocity; and before this, the weights raised are nearly as the squares of the velocities with which the wheel is struck. To compare which, the eighth column is added, and is found by proportion. For as the seventh is only comparative, if we say, beginning with the maximum, Experiment 4th, Table 9th, As

72.9, the real effect, is to 119.16, the relative
effect, so is 72.8 to 118.9, the relative, &c.; by
which process, the eighth column is formed
from the experiments, in order to compare
them with the theory.

ON UNIFORM VELOCITY.

A wheel turned by the weight or impulse of
water has its velocity increased till the resistance becomes equal to the accelerating force.
The resistance may be of various kinds; and
in many cases it may be not only difficult, but
impossible to bring it into a computation. See
page 155. In the present observations, I shall
consider the resistance as arising from the
weight of the wheel, and a weight to be raised
by a cord fixed to its axle.

The following experiments were made with
a bucket-wheel 20 inches in diameter, and supported on friction-wheels: the circumference
of its axle 9 inches; the whole weight of the
wheel and axle 8lb.; the centre of gyration is
at 8.4 inches; and the resistance, at the distance of 10 inches, will be about 92 ounces.

TABLE X.

Number.	Weight raised, in ounces.	Resistance of the wheel.	The whole resistance.	Time of rising 5 feet.	Velocity of the wheel, in feet, per second.	Velocity, by computation.	Ounces of water on the wheel.
	m	q	p	t	v		
1	4.5	3.5 oz	8 oz.	$13\frac{1}{2}''$	2.48	2.89	2.68
2	8.5	2.35	10.85	$16\frac{1}{2}$	2.03	2.12	3.28
3	12.5	1.54	14.04	$21\frac{1}{4}$	1.55	1.63	4.3
4	16.5	.98	17.48	27	1.24	1.30	5.37
5	20.5	.7	21.2	32	1.05	1.07	6.35
6	24.5	.52	25.02	37	.90	.92	7.4

In the first experiment, the water-wheel revolves 6.7 times in 13½ seconds, and raises 4¼
ounces through 5 feet; the velocity of the circumference is 2.48 feet per second.

In the second experiment, the wheel makes
the same number of turns in 16⅛ seconds, and
raises 8½ ounces through the same space, &c.

We are next to inquire, what weights applied
to the axle would turn the wheel, and descend
through 5 feet in 13½, 16⅛, 21¾, &c. seconds
(See Theo. 4, page 125). We have given,

F = 5 feet, the space passed over by the weight.
m = the weight suspended from the axle.
w = 92 oz., the weight of the wheel multiplied
    by the square of the centre of gyration,
    and divided by the square of its radius.
s = 10 inches, the radius of the wheel.
r = 1.42 inches, the radius of the axle.
t = the time of descent.
d = 16, the feet fallen through by gravity, in
    one second.

Then we have the weight $m = \dfrac{Fws^2}{dr^2t^2 - Fr^2} =$
7.02 oz., when t = 13½ seconds; but when
t = 16⅛, m = 4.7 oz. &c. Now, if 7 oz. falls
through 5 feet with an accelerated velocity, by
the laws of falling bodies, it would, at the end

of the time, have acquired a velocity which, if equable, would carry it over 10 feet in the same time. And the whole effect is equal to what half the weight would produce in passing over the same space in the same time, with an uniform velocity. Half the weights found by the above theorem are placed in the second column, under q, which being added to the weight raised, gives the whole resistance, as in the next column.

To find the Quantity of Water upon the Wheel.

If the quantity delivered in one second be multiplied by the seconds in which the wheel makes half a revolution, the product will be the whole quantity upon the wheel.

Or, if half the circumference, in feet, is multiplied by the quantity per second, and divided by the velocity, the quotient will be the quantity upon the wheel.

Let q = the quantity per second.

$\quad$ c = half the circumference.

$\quad$ v = the velocity of the circumference.

$\quad$ $m + q = p$ = the sum of the weight and resistance.

Then will $\frac{qc}{v}$ = the quantity upon the wheel, which quantity will vary with the velocity.

z

$\frac{pr}{s}$ = the power applied at the circumference, which will balance $m + q$, when in motion. Hence, $\frac{Qc}{v} - \frac{pr}{s} = \frac{Qcs - vpr}{vs}$ = the efficacious moving power; and the resistance is $\frac{Qc}{v} + w +$ $\frac{pr^2}{s^2} = \frac{Qcs^2 + wvs^2 + pvr^2}{vs^2}$; by which divide the moving power, and we have $\frac{Qcs^2 - vp\cdot s}{Qcs^2 + uvs^2 + pvr^2}$ for the accelerating force, which, when the motion becomes uniform, must be equal to the retarding force, which will be expressed by $\frac{pr}{s} \div w$ $+ \frac{pr^2}{s^2} = \frac{prs}{s^2w + r^2p}$; from which we find $v =$

$$\frac{Qcs}{2pr} \times \frac{s^2w + pr^2 - prs}{s^2w + pr^2} = \frac{Qcs}{2pr} \times \overline{1 - \frac{prs}{ws^2 + pr^2}}.$$

The value of v found by the theory, may next be compared with what it is in the experiment, in which we have given $Q = 2\frac{2}{3}$ oz., $c = 2\frac{1}{2}$ feet, the rest of the notation as above, except p, which is different in every experiment.

In the first experiment $p = 8$, and $\frac{Qcs}{2pr} =$ $\frac{66.66}{22.72} = 2.934$; and $1 - \frac{prs}{ws^2 + pr^2} = .987$, which multiplied by 2.934, gives $2.897 = v$.

In the second, $p = 10.85$, and $v = 2.12$, as in the table; in which it appears, that the velocity found by computation is, in every experiment, something too much: but the difference is less than could be expected, when we consider the degree of accuracy which is required in the quantity of water, the weight, diameter, inertia, friction, &c.

When half the buckets are equally loaded with water, and the wheel at rest; if the whole weight of the water is multiplied by .6366, the product is the weight which, if suspended from the end of the opposite horizontal radius, would balance the whole. But, as near as I can discover by experiment, when the wheel is in motion, the effect is nearly the same as if all the water in the buckets gravitated on the end of the horizontal diameter. If this is true, it may be owing, in part, to the velocity which the water has when thrown upon the wheel, and to the centrifugal force, by which it endeavours to move in a tangent to the wheel. But, in making computations, to be on the safe side, it will be better to multiply the quantity by .6366.

When the resistance is irregular, as in forges, tilts, saw-mills, rolling-mills, &c. it becomes necessary to add a fly, in order to preserve, as near as may be, an equable motion, and to avoid sudden shocks in the work.

Where there is convenience, a large fly ought to be preferred. For if a fly of 10, and another of 40 feet diameter, revolve in the same time, the smaller must be 4 times as heavy as the other to produce the same effect.

In page 144, 5lb. is raised 6 feet in 142 seconds, or through 21 inches, the diameter of the wheel, in 42 seconds, in which time 7lb. of water descends through the same space; hence, two-sevenths of the whole is employed in overcoming the friction, resistance, &c., and the power is to the effect as 7 to 5.

In the sixth experiment of the last Table, 24.5 oz. is raised through 5 feet in 37 seconds, or through the diameter of the wheel in 12.95 seconds, in which time 34.5328 oz. of water descends through the same space; and As 24.5 : 34.5328 :: 5 : 7.03, so is the effect to the power, the same as above, nearly. If more water is applied, the velocity is increased; but in the same wheel, the ratio of the power to the effect will always be nearly the same, at a maximum.

Index.

A.

AIR, its resistance, 125, 127.
Angular velocity, the same under what circumstance, 36.
Application of water to wheels, 134, 138.

B.

Buckets and floats compared, 160.

C.

Centre of gravity, what, 46—problems concerning, 32, 35, 48.
———— of gyration, 43—problems concerning, 44, 45.
———— of oscillation, 49, 67.
Centrifugal and centripetal force, what, 2—problems concerning, 7, 19,—when equal in different wheels, 9.
Circular motion, a machine for making experiments on, 122—experiments thereon, 136, 142,—quantity produced by a given power, 37.

D.

Different quantities of water, their effects, 145, 146.

E.

Experiments on the velocity of effluent water, 100, 105.
——————— on the impulsive force of water, 114, 115.
——————— on emptying of vessels, 116—with syphons, 118.
——————— on the velocity with which water ascends into a vacuum, 119.
——————— on circular motion, 126, 131.

INDEX.

Experiments on the application of water, 134.

—————— on the velocity of wheels, 138—with different quantities of water, 145.

—————— on the size of wheels, 148.

—————— on the fall of water, 152.

—————— on impulse and gra ity, 156.

F.

Floats and buckets compared, 160.

Fall of water, 152, 156.

Force of a moving bar, how computed, 43.

Friction, 155.

G.

Grinding stones, their velocity, &c. 9.

Gyration, centre of, how found, 43, 44, 64.

I.

Inclined plane, ascent of bodies thereon, 21.

Initial velocity of a machine, &c. 20, 37, 57, 59, 60, 63.

Impulse of water, 112—experiments thereon, 114.

Impulse and gravity compared, 156, 159.

L.

Lever, its greatest effect, demonstrated, 24—problems concerning, 32, 33, &c.

M.

Machine for circular motion, 122.

————— for measuring the force of a stream, 114.

Methods of computing the force of moving bodies, 24.

Momentum of a moving bar, 33.

O.

Oscillation, centre of, 49, 55, 67.

P.

Percussion, centre of, what, 49, &c.

INDEX.

R.

Resistance of the air, 125, 127.
————— of machines, 155.

S.

Steam engines, theory of, 85, 95.
Syphons, experiments with, 118.

T.

Time of emptying vessels, 116.

V.

Vacuum made by water, 121.
Velocity of water through thin plates, 100 —through cones, 101—how measured, 99—through notches, 104, 109, —into a vacuum, 119, 120.
————— of water-wheels, 107—when different quantities of water are applied, 18, 19, 145—quantity communicated to a wheel by a given power, 36, 42, 58, 67.
————— angular, the same under what circumstances, 36.

W.

Water, how best applied to wheels, 134, 138—effects of different quantities, 145, 146—its velocity through syphons, 118—its velocity obtained by emptying vessels, 116.
Wheels driven by impulse, their maximum, 30, 162—what magnitude to prefer, 148, 151—their velocity, which best, 138, 145.
Wheel and axle, its maximum, at the end of a given time, 23, 67—at the end of a given space, 74, 78, 79.

FINIS.

PRINTED BY G. F. HARRIS'S WIDOW AND BROTHERS,
WATER-STREET, LIVERPOOL.

Printed in the United States
By Bookmasters